The Complete Guide to Defense in Depth

Learn to identify, mitigate, and prevent cyber threats with a dynamic, layered defense approach

Akash Mukherjee

The Complete Guide to Defense in Depth

Group Product Manager: Dhruv Kataria
Publishing Product Manager: Prachi Sawant
Book Project Manager: Srinidhi Ram
Senior Editor: Sayali Pingale
Technical Editor: Nithik Cheruvakodan
Copy Editor: Safis Editing
Indexer: Pratik Shirodkar
Production Designer: Ponraj Dhandapani
Senior DevRel Marketing Executive: Marylou De Mello

First published: July 2024

Production reference: 1100724

Published by Packt Publishing Ltd.

Grosvenor House
11 St Paul's Square
Birmingham
B3 1RB, UK

ISBN 978-1-83546-826-5

www.packtpub.com

Embarking on the journey of writing this book has been a remarkable experience, filled with challenges and rewards, and I am deeply grateful to all those who have contributed to its realization. I would like to thank my loving wife, Roop, and my golden boy, Champ, for their unwavering support throughout this endeavor. To my amazing parents, your endless sacrifices and boundless love have shaped me into the person I am today. I am forever grateful for your guidance, wisdom, and unconditional support.

– Akash Mukherjee

Foreword

In some ways, the defenders' job is getting easier, relative to the threats that we faced in the 2000s and 2010s. In modern enterprises, we have available to us a range of endpoint protection, anomaly detection, data loss prevention systems, secure operating systems, hardened cloud platforms, supply chain analysis, automated defect analysis, and counter-abuse technologies to bring to bear on the attackers that face our systems. Though some of these options may be expensive, they are well understood and available.

In other ways, the threat ecosystem is getting more sophisticated and more bespoke with no clear strategy for our defenders. We now regularly discuss the proliferation of weaponized 0-days, critical infrastructure, and ransomware attacks, the utilization of AI to automate the exploitation of low-hanging vulnerabilities in enterprise systems and people, and increased visibility of nation-state-level attacks in headlines. A book such as this one is exactly what defenders need.

When faced with this evolving ecosystem, how does one take all of the pieces that are available – from best practices and third parties—and integrate them into a holistic layered defense strategy? At the level of an individual security engineer, one must now consider the full range of all vulnerability exploitation that may occur in the system that is being analyzed: zero-trust is the new default. It's no longer the case that one can assume that nation-states are not interested in the software that you are securing, or more likely, that they may not have infiltrated your third-party dependency graph.

In this book, Akash brings you, dear reader, well-wrought experience from the frontier of defense, from the heart of the most sophisticated cybersecurity teams at the most advanced tech companies –where I was honored to work with him! Against the most advanced attackers, the strategy is made accessible with real-world case studies built on hard engineering problems – not lofty enterprise-laden jargon. For the defender looking to increase their chances of succeeding against advanced detectors, this book offers guideposts and actionable advice.

One is loath to make predictions about the future, but it seems inevitable that the Red Queen hypothesis dynamics as applied to cybersecurity will continue to accelerate in ways in which traditional defense strategies will no longer be able to keep up. Threat actors have at their disposal new tools that can fully automate parts of the attack chain and, in a few years with the proliferation of AI agents, even the entire attack chain. In this environment, every defender needs a guide, such as this book, to help them understand how to build a resilient enterprise or system that plans for, resists, and mitigates proliferating exploitation before the impact of this change is felt.

Jason D. Clinton

Chief Information Security Officer, Anthropic

Contributors

About the author

Akash Mukherjee is a security enthusiast and a leader with experience setting up and executing security strategies at large tech companies. He is currently a security leader at Apple AIML. He was previously a security lead at Google, leading the insider risk program and supply chain security efforts at Google Chrome. During his time at Google, Akash was also a course lead and subject matter expert for the Google Cybersecurity Certificate course. He has been at the forefront of the emerging threat landscape and has led the development of novel security strategies and frameworks. Akash was one of the co-developers of the open-source **Supply-chain Levels for Software Artifacts** (**SLSA**) framework.

He is based in the Silicon Valley area in the US, and he holds a bachelor of technology degree from the Indian Institute of Technology, B.H.U., India, and a master's degree in cyber security from the University of Southern California, USA.

I am immensely grateful to those who have stood by me and offered unwavering support, especially my wife, Roop, and my parents and friends.

I would also like to extend my appreciation to the Packt team for their help in refining the manuscript and improving its quality.

About the reviewers

Arun Kumar has extensive experience in cyber security and telecommunications. He is an active member of EC-Council, ISC2, and PMI. He has led cybersecurity teams for drone defense, medical devices, banking, finance, and insurance companies. He started his career in telecommunications engineering, moving to project teams and eventually leading engineering projects. In his spare time, he enjoys volunteering at PMI and ISC2.

Peter Bagley retired from the US Army as an information system analyst after 21 years, and now he's the CIO of B&B Cyber Solutions LLC and has worked as a senior cyber engineer/ISSM, supporting enterprise security and vulnerability management using NIST **Risk Management Framework** (**RMF**), NIST SP 800-53, **Cybersecurity Framework** (**CSF**), and **Cybersecurity Maturity Model Certification** (**CMMC**) for NIST SP 800-171. He has over 36 years of IT/cyber experience and over 30 years of teaching. Currently, he is a cybersecurity professor for St. Petersburg College and a cyber training consultant in the Tampa Bay area. He holds an MS in information systems from the University of Maryland, and several industry certifications, including CISSP, CMMC-RP, ISO-27001, CEH, and CHFI.

Gursev Singh is an accomplished, results-oriented cybersecurity expert with over 16 years of experience in infrastructure design and enterprise security. He has a proven track record of success in leading and managing client projects in areas such as public cloud security, SIEM data protection, infrastructure security, and cyber threat and vulnerability management. Gursev has worked with several leading organizations, including Deloitte, Esri, VMware, and Quest Software. In his current role at Google, he independently leads and manages client projects within the Cyber Risk Management Services service offering. He is currently pursuing a master of science in cyber security operations and leadership at the University of San Diego.

I would like to extend my heartfelt thanks to my Mom and Dad for always believing in me and giving me the opportunity to pursue bigger things in life. I am also deeply grateful to my loving wife, Kanchan, and my kids, Arsh and Gyanve, for their unwavering support throughout this journey. Your love and encouragement have been my guiding light.

Table of Contents

3

Building a Framework for Layered Security   63

Part 2: Building a Layered Security Strategy – Thinking Like an Attacker

4

5

Part 3: Adapting and Evolving with Defense in Depth – The Threat Landscape

9

The Human Factor – Security Awareness and Training   211

10

Defense in Depth – A Living, Breathing Approach to Security   239

Preface

Let's start with a question. In the face of modern adversaries, can a system be deemed secure if it uses the latest technology at the edge? Fundamentally, there are a couple of issues in this question. First, there are no "perfectly" safe systems, only safer ones. Second, security is not about protecting the perimeters anymore; attackers are looking for gaps in our design from all directions.

Defense in Depth is a security design principle that layers security controls to protect, acknowledges the inevitability of failures, and focuses on resilience to create a formidable barrier against the modern threat landscape. Recent attacks such as the SolarWinds attack taught us that protecting the interfaces of a system is not enough; security needs to be part of every phase of the software development life cycle. If we break down security practices in organizations, they can be broadly categorized as follows:

- Application or product security, sometimes platform security
- Enterprise, corporate, and infrastructure security
- Security governance, policy, and compliance

There are plenty of good resources that cover these topics individually. However, successfully designing, building, and maintaining robust security systems is much more complex than a random mix of these pillars. As attackers grow ever more sophisticated, using AI and automated tools, Defense in Depth provides a structured, proactive framework for building resilient systems designed to withstand the onslaught.

As we become more reliant on the digital ecosystem, **security by default** will become increasingly relevant. To be able to secure software against advanced cyber threats, one needs a holistic understanding of the individual pieces and their interplay. In this book, I aim to provide a comprehensive overview deeply rooted in security-first principles. I will guide you through real-world attacks to help you build a mental map and a framework that can withstand advanced threats.

Defense in Depth is in the spotlight in every critical security role today. The escalating frequency and sophistication of cyberattacks are only going to drive the surge. High-profile breaches have exposed the futility of relying solely on prevention. Defense in Depth acknowledges this, providing a practical framework for resilience. It emphasizes layered protection, continuous monitoring, and strategies to limit the damage caused by successful attacks.

As demands grow, Defense in Depth is going to be a crucial skill for every security professional and it will have faster growth opportunities.

Who this book is for

Security is everyone's responsibility, so we are targeting a wide audience. This book is designed for anyone working in the cybersecurity field, including security analysts, security engineers, security architects, and security managers, who can all benefit from reading this book.

Three main personas who are the primary target audience for this content are as follows:

- **Security leads**: Leaders who design and architect security systems and strategies for organizations.
- **Security developers**: Individuals who design, implement, and maintain security controls and work with developers for enforcement. This book will provide a comprehensive guide to help them grow in their career to emerge as security leaders.
- **Business leads**: Leaders who drive business outcomes and make directional decisions about an organization's roadmap. This book will provide real-world case studies to make informed business decisions and encourage them to include security as a core value of the company.

Throughout the book, when we say "you," we mean you, the reader, irrespective of your role or experience. We believe security is a collective journey and everyone plays an equal part in it.

What this book covers

Chapter 1, Navigating Risk, Classifying Assets, and Unveiling Threats, serves as a comprehensive introduction to the fundamental principles of security. By adopting a risk-based approach, the chapter provides you with an in-depth examination of asset classification and the various categories of threat actors, along with their underlying motivations.

Chapter 2, Practical Guide to Defense in Depth, builds upon the risk-based approach to security strategies and lays the foundation for Defense in Depth. It places significant emphasis on various security domains and the diverse range of controls within them. This chapter introduces primary components in a layered security design with a glimpse of real-world applicability.

Chapter 3, Building a Framework for Layered Security, reinforces the core principles of security and deepens the understanding of defense in depth, laying the foundation for crafting resilient security strategies. It emphasizes the critical role of introducing and implementing security policies to govern large-scale changes within organizations.

Chapter 4, Understanding the Attacker Mindset, focuses on types of threat actors and common tactics used by them. It covers the importance of understanding the adversaries to build a strong security strategy.

Chapter 5, Uncovering Weak Points through an Adversarial Lens, delves into the intricacies of adopting an attacker's perspective to fortify defense systems. Based on the unique threat landscape for every organization, this chapter demonstrates how to craft tailored defense programs by profiling these risks.

Chapter 6, *Mapping Attack Vectors and Gaining an Edge*, focuses on drawing the line between common threats that organizations face and the attacker mindset to build a formidable security strategy. A lot of attention is paid to practical defense in depth security controls to give you the ability to understand the common attacks and be able to create a layered security posture.

Chapter 7, *Building a Proactive Layered Defense Strategy*, provides an overview of designing defense in depth using proactive, attacker-focused strategies. You will learn how to characterize different security mechanisms into buckets and apply them to appropriate situations.

Chapter 8, *Understanding Emerging Threats and Defense in Depth*, delves a little deeper into adaptive defense strategies based on evolving threat vectors. A lot of attention is paid to the effectiveness of a defense in depth approach against emerging threats and how to utilize advanced technologies as core components in defense systems.

Chapter 9, *The Human Factor – Security Awareness and Training*, introduces one of the most important gaps in today's security world: humans. Building on top of zero trust principles, this chapter puts the focus on security as a chain and intrinsic weakness by design. It discusses the idea of leaving humans out of the loop to increase the robustness of security and also touches on the concept of reliability.

Chapter 10, *Defense in Depth – A Living, Breathing Approach to Security*, provides an overview of the inevitability of defense in depth in modern security models. Introducing the Secure Software Development Framework, this chapter demonstrates how to build a security program with defense in depth at the center of it. You will learn why defense in depth is the only way to think about building security strategies.

To get the most out of this book

This book targets a wider audience and does not assume any prior knowledge. It builds on top of fundamental security concepts. Having some exposure to security challenges might come in handy for the real-world case studies presented; however, we have added further reading to solidify some of these gaps as you progress through the book.

Conventions used

There are a number of text conventions used throughout this book.

Bold: Indicates a new term, an important word, or words that you see onscreen. For instance, words in menus or dialog boxes appear in **bold**. Here is an example: By understanding the **tactics, techniques, and procedures** (**TTPs**) employed by adversaries, organizations can implement countermeasures and mitigate potential attacks.

> **Tips or important notes**
> Appear like this.

Get in touch

Feedback from our readers is always welcome.

General feedback: If you have questions about any aspect of this book, email us at customercare@packtpub.com and mention the book title in the subject of your message.

Errata: Although we have taken every care to ensure the accuracy of our content, mistakes do happen. If you have found a mistake in this book, we would be grateful if you would report this to us. Please visit www.packtpub.com/support/errata and fill in the form.

Piracy: If you come across any illegal copies of our works in any form on the internet, we would be grateful if you would provide us with the location address or website name. Please contact us at copyright@packt.com with a link to the material.

If you are interested in becoming an author: If there is a topic that you have expertise in and you are interested in either writing or contributing to a book, please visit authors.packtpub.com.

Share Your Thoughts

Once you've read *The Complete Guide to Defense in Depth*, we'd love to hear your thoughts! Scan the QR code below to go straight to the Amazon review page for this book and share your feedback.

https://packt.link/r/1835468268

Your review is important to us and the tech community and will help us make sure we're delivering excellent quality content.

Download a free PDF copy of this book

Thanks for purchasing this book!

Do you like to read on the go but are unable to carry your print books everywhere?

Is your eBook purchase not compatible with the device of your choice?

Don't worry, now with every Packt book you get a DRM-free PDF version of that book at no cost.

Read anywhere, any place, on any device. Search, copy, and paste code from your favorite technical books directly into your application.

The perks don't stop there, you can get exclusive access to discounts, newsletters, and great free content in your inbox daily

Follow these simple steps to get the benefits:

1. Scan the QR code or visit the link below

https://packt.link/free-ebook/9781835468265

2. Submit your proof of purchase
3. That's it! We'll send your free PDF and other benefits to your email directly

Part 1: Understanding Defense in Depth – The Core Principle

In this part, we focus on building a strong foundation of security, establishing the core theme of the book by introducing a risk-based approach to security. We'll begin by demystifying the world of cyber risk, helping you identify what assets are most valuable and the threats they face. You'll learn the fundamentals of Defense in Depth, and how it translates into practical strategies. Finally, we'll guide you through creating a security framework that combines layers of protection, tailored to your unique needs. Approach this part as learning or refreshing concepts around the building blocks that make up Defense in Depth.

This part has the following chapters:

- *Chapter 1, Navigating Risk, Classifying Assets, and Unveiling Threats*
- *Chapter 2, Practical Guide to Defense in Depth*
- *Chapter 3, Building a Framework for Layered Security*

Navigating Risk, Classifying Assets, and Unveiling Threats

The realm of security demands a dynamic approach that encompasses risk management, asset classification, and **threat intelligence** (**TI**). In this comprehensive exploration, we embark on a journey through the foundational building blocks of cybersecurity. In this ever-evolving landscape, the chapter serves as a compass, guiding through the complex web of risks.

Risk is omnipresent in the digital realm, and understanding its intricacies is paramount. Effective security strategies hinge on a thorough understanding of the inherent risks associated with an organization's operations. Identifying and assessing these risks allows for the prioritization of security measures and the allocation of resources accordingly.

Asset classification forms the bedrock of any strategic defense measures. By meticulously classifying assets, organizations can determine the criticality of each asset and implement appropriate safeguards. A comprehensive classification process ensures that the most valuable assets receive the necessary level of protection. This helps organizations better understand their exposure, effectively calculate risks, and allocate resources to safeguard against threats.

Unveiling the ever-changing landscape of threats is crucial for proactive defense. Continuous monitoring and analysis of TI enable organizations to stay ahead of emerging threats and adapt their security posture accordingly. By understanding the **tactics, techniques, and procedures** (**TTPs**) employed by adversaries, organizations can implement countermeasures and mitigate potential attacks.

As we walk through the intricate interplay of risks, assets, and threats, this chapter aims to lay the groundwork for a proactive and strategic approach to security in general. By the end of this chapter, you will have developed a comprehensive framework for categorizing modern threats and effectively navigating the complex cyber-security landscape.

In this chapter, we're going to cover the following main topics:

- Foundations of security principles
- Risk-based approach to security

- Identifying threat actors and understanding their motivations
- Security through the ages

Let's get started!

Foundations of security principles

Cybersecurity is becoming increasingly complex. Due to its complexity, security is one of the major challenges organizations face. To fully understand the inherent nature of these challenges, we peek into the evolution of this vast, dynamic field. Positioned at the crossroads of humans and systems, the security landscape is filled with technical and procedural controls.

Today, security is a fundamental concern for every organization. Strong security practices are essential to protect against unauthorized access, data breaches, and other cyberattacks. The foundations of security principles provide a framework for understanding and implementing effective security measures.

To understand the inception of the field of cybersecurity, let's draw a rough timeline. The first concept of a digital computer was originated by Charles Babbage in 1822. But the first computer was not built until the 20th century. On the other hand, the first-ever "virus" called "the Creeper" was built by Bob Thomas in the 1970s, the era of the **Advanced Research Projects Agency Network** (**ARPANET**). While the concepts of security still applied to all computers, security's prominence rose alongside the modern surge in technological advancements. It is crucial to understand this relevance to build a robust security strategy in the modern world. Let's take a deeper dive into a brief history of information security.

Brief history of information security

The realm of information security has evolved alongside the development of information technology, with its roots tracing back to ancient civilizations' methods of safeguarding sensitive information. As early as 1900 BC, the Egyptians employed sophisticated encryption techniques to protect their valuable papyri and scrolls. The advent of cryptography, the study of secure communication techniques, played a pivotal role in safeguarding information throughout history.

In the mid-20th century, the development of early computers raised concerns about securing data and systems. The 1970s marked a turning point in information security as computers became increasingly prevalent and interconnected. The creation of ARPANET, the precursor to the modern internet, brought with it new cybersecurity challenges. The first computer virus, the Creeper virus, emerged in 1971, demonstrating the vulnerability of interconnected systems to malicious attacks. Notable milestones include the creation of the first computer-specific security model by Bell-LaPadula in the 1970s, laying the groundwork for access control in computer systems.

The evolution of the internet and networking further propelled the need for robust security measures. With the emergence of TCP/IP protocols in the 1980s, concerns about data integrity, confidentiality, and network security became paramount. The 1980s and 1990s saw a proliferation of cyber threats, including the rise of hacking, malware, and **denial-of-service** (**DoS**) attacks. The Morris worm, unleashed in 1988, wreaked havoc on internet infrastructure, highlighting the potential for large-scale cyberattacks. The advent of the World Wide Web in the 1990s expanded the attack surface, exposing businesses and individuals to new security risks.

The 20th century witnessed a surge in information security advancements driven by the emergence of computing and the rise of cyber threats. During World War II, the breaking of German military codes, such as Enigma, by Allied forces exemplified the importance of information security in national defense and strategic operations. The Cold War era saw a heightened focus on espionage and counterintelligence, further emphasizing the need for robust information protection measures.

The 21st century has witnessed an explosion of data and the rapid adoption of cloud computing, mobile devices, and the **Internet of Things (IoT)**, creating an ever-evolving cybersecurity landscape. The threat landscape has become increasingly sophisticated, with attackers employing advanced techniques such as social engineering, zero-day exploits, and targeted attacks. **Artificial intelligence (AI)** and **machine learning (ML)** are transforming the field, enabling both attackers and defenders to develop more sophisticated tools and techniques.

The evolution of information security is characterized by a constant cat-and-mouse game between security professionals and malicious actors. Today, information security has become an indispensable aspect of modern society, safeguarding the confidentiality, integrity, and availability of information assets that are essential for individuals, organizations, and nations. As technology continues to evolve, the importance of information security remains paramount, demanding continuous vigilance, innovation, and collaboration to protect the digital world.

The CIA Triad – Confidentiality, integrity, and availability

The CIA Triad, encompassing *confidentiality*, *integrity*, and *availability*, stands as the cornerstone of information security. Comprehending these fundamental principles is paramount, as any security breach can be traced back to a violation of one or more of these core properties. The CIA Triad serves as a robust framework for understanding the foundations of security, providing a comprehensive lens through which to assess and address security risks:

Figure 1.1 – The CIA Triad

To delve deeper into the CIA Triad's significance, let's examine each principle in detail:

- **Confidentiality**: Confidentiality safeguards the privacy of information, ensuring that it is only accessible to authorized individuals. A breach of confidentiality occurs when unauthorized individuals gain access to sensitive data, potentially leading to identity theft, financial fraud, or reputational damage.

- **Integrity**: Integrity ensures the accuracy and consistency of information, preventing unauthorized modifications or alterations. A breach of integrity occurs when information is modified or corrupted, potentially leading to erroneous decisions, disrupted operations, or legal repercussions.

- **Availability**: Availability ensures that authorized individuals have timely and reliable access to information and systems. A breach of availability occurs when systems become inaccessible or inoperable, potentially hindering critical business processes, disrupting customer service, or causing financial losses.

By upholding the CIA Triad, organizations can establish a strong foundation for protecting their valuable information assets and ensuring the continuity of their operations. The CIA Triad serves as a timeless and adaptable framework, guiding organizations in navigating the ever-evolving cybersecurity landscape.

Security standards, policies, and guidelines

The foundation of a robust cybersecurity posture lies in the establishment of clear and comprehensive security standards, policies, and guidelines. These frameworks provide a structured approach to safeguarding sensitive information and ensuring the confidentiality, integrity, and availability of critical assets.

Security standards

Security standards are formalized sets of rules or best practices developed by recognized organizations, such as the **International Organization for Standardization** (**ISO**) and the **National Institute of Standards and Technology** (**NIST**). They serve as benchmarks for organizations to assess and implement effective security controls, fostering consistency and interoperability within the cybersecurity landscape.

The benefits of having security standards include the following:

- **Establishing a common baseline**: Adherence to recognized standards ensures that organizations meet a minimum level of security effectiveness.

- **Providing implementation guidance**: Standards outline specific security controls and practices to address various security risks.

- **Facilitating compliance**: Standards align with regulatory requirements, simplifying compliance efforts.

- **Enhancing collaboration**: Common standards enable organizations to share information and collaborate on security initiatives.

Security policies

Security policies are high-level directives that articulate an organization's overall security posture and objectives. They define the organization's stance on security matters, providing a framework for decision-making and ensuring that security is embedded into organizational culture.

A security policy must have the following key characteristics:

- **Scope**: Define the boundaries of the policy and the entities it applies to.
- **Purpose**: Clearly state the overall security goals and objectives of the policy.
- **Requirements**: Outline specific security controls and practices that must be implemented.
- **Responsibilities**: Assign ownership and accountability for implementing and enforcing the policy.
- **Enforcement**: Define mechanisms for monitoring, enforcing, and addressing policy violations.

Security guidelines

Security guidelines provide detailed, practical guidance on how to implement and maintain security controls. They supplement security policies by offering specific recommendations and best practices, tailored to specific security domains or technologies.

Some crucial characteristics of security guidelines are the following:

- **Actionable**: Provide step-by-step instructions for implementing security controls.
- **Technical**: Address specific technical configurations.
- **Domain-specific**: Focus on specific areas of security, such as network security or data security.
- **Freshness**: Regularly review and update guidelines to reflect evolving threats and technologies.

In a nutshell, security standards, policies, and guidelines work in tandem to establish a comprehensive security framework. Standards provide a foundation of best practices, policies set the overall direction, and guidelines offer practical implementation guidance. This synergistic approach ensures that organizations have a well-defined and effective approach to ever-evolving threats and attack strategies in the field of cybersecurity.

Evolution of cyber threats and attack strategies

The landscape of cyber threats has undergone a dramatic transformation over time, mirroring the rapid advancements in technology and our increasing reliance on digital systems. Early threats were primarily focused on technical vulnerabilities and involved simple malware or DoS attacks. Over time, the sophistication and scope of cyber threats have steadily escalated, necessitating continuous adaptation and innovation in security strategies. Later in this book, we'll delve into attacker motivations, which significantly influence various attack techniques. For now, let's explore key highlights from the past few decades.

Early days – Simple malware and DoS attacks (1990s)

The early days of the internet saw basic malware programs designed to disrupt or disable systems. These attacks were often opportunistic, exploiting generic vulnerabilities and lacking significant strategic intent. DoS attacks aimed to overwhelm systems with traffic, making them inaccessible to legitimate users. Examples include the Morris worm in 1988 and the Melissa virus in 1999.

Rise of targeted attacks and zero-day exploits (2000s)

With the proliferation of sensitive data online, the 2000s saw the emergence of more sophisticated threats. Hackers developed targeted attacks, leveraging social engineering and zero-day exploits (unknown vulnerabilities) to gain access to critical systems and steal valuable information. Notable examples include the Titan Rain campaign targeting US government agencies and the Stuxnet attack on Iran's nuclear program.

Ransomware and supply chain attacks (2010s)

The past decade witnessed new and dangerous forms of cyber threats. Ransomware attacks, encrypting data and demanding payment for decryption, became a major concern, impacting businesses, government agencies, and individuals. Supply chain attacks also emerged, targeting third-party vendors to gain access to customers' sensitive information. Examples include the WannaCry ransomware outbreak in 2017 and the SolarWinds attack in 2020, compromising numerous government agencies and private companies.

Recent threats – Heartbleed, Apache Struts, and Heroku attack (2020s)

The evolution continues with increasingly complex and impactful attacks. In 2014, the Heartbleed bug exposed sensitive information on major websites. The 2017 Apache Struts vulnerability allowed hackers to remotely execute code on vulnerable servers. In 2023, the Heroku attack exposed data belonging to thousands of developers and their clients. These highlight the need for robust security across the entire digital ecosystem.

Continuous adaptation – The need for innovation

The ever-changing threat landscape demands continuous adaptation and innovation in cybersecurity. Organizations must constantly update their defenses, stay informed of emerging threats, and adopt new technologies and strategies to stay ahead of adversaries. While the core principles of confidentiality, integrity, and availability remain paramount, tactics for upholding them must evolve alongside the evolving threats. In addition to CIA principles, some security controls form the bedrock of evolving defense strategies.

Let's cover some foundational security controls in the next section.

Security controls

Security controls serve as the linchpin in establishing a robust security posture within any organization. This section explores the pivotal role of security controls as the bedrock of an effective security foundation. Security controls encompass a diverse array of measures, methodologies, and technologies designed to safeguard information, systems, and networks from potential threats and vulnerabilities. These controls operate across various layers of an organization's infrastructure, aiming to mitigate risks, enforce policies, and ensure the confidentiality, integrity, and availability of critical assets.

Types of security controls

Security controls are classified into different categories based on their primary objectives and functionalities. They span across preventive, detective, corrective, and deterrent measures. Preventive controls aim to stop security incidents before they occur, while detective controls focus on identifying and alerting about potential security breaches. Corrective controls come into play after an incident, helping to restore systems and mitigate damage, while deterrent controls aim to dissuade attackers from targeting systems or assets.

Common security controls

A variety of security controls exist, each serving specific purposes within an organization's security framework. These include but are not limited to access controls, encryption mechanisms, firewalls, **intrusion detection systems** (**IDSs**), antivirus software, and security policies and procedures. These controls collectively form a **Defense in Depth** (**DiD**) strategy, working in tandem to fortify the organization's resilience against a myriad of cyber threats.

IAAA – The four pillars of secure access management

In the realm of cybersecurity, the CIA triad stands as the cornerstone of information security. However, another fundamental framework, IAAA, plays a crucial role in ensuring secure access management. Let's delve into IAAA and understand how it safeguards valuable information:

- **Identity (I):**
    - **The foundation**: IAAA begins with identity, establishing a clear and accurate understanding of users within the system. This involves assigning **unique identifiers** (**UIDs**) and implementing robust user management practices. Effective **identity management** (**IdM**) ensures users are who they claim to be, preventing unauthorized access and impersonation.
    - **Examples**: User IDs, email addresses, usernames, digital certificates.

- **Authentication (A):**

  - **Verifying claims**: Authentication verifies user identities through credential checks. This typically involves requiring passwords, biometrics, or **multi-factor authentication (MFA)** to confirm user claims. MFA adds an extra layer of security, further reducing the risk of unauthorized access.

  - **Examples**: Passwords, PINs, biometric scans (fingerprint, facial recognition), **one-time passwords (OTPs)**.

- **Authorization (A):**

  - **Defining access privileges**: Once a user's identity is verified, authorization defines their access privileges within the system. This involves assigning specific permissions based on individual roles and responsibilities. Authorization ensures users only have access to resources and information they need, minimizing the risk of data breaches or misuse.

  - **Examples: Access control lists (ACLs), role-based access control (RBAC), attribute-based access control (ABAC)**.

- **Audit (A):**

  - **Monitoring and accountability**: Auditing plays a vital role in monitoring user activity and identifying potential security incidents. It involves logging and analyzing user actions to ensure compliance with security policies and detect suspicious behavior. Audit trails provide valuable evidence for investigations and enable organizations to take corrective actions.

  - **Examples**: Activity logs, session logs, access logs, audit reports.

Let's go over an example to solidify our understanding of these concepts.

In a production web server, we will talk about two users: *Alice* and *Bob*. Alice is an admin user, whereas Bob is a regular web application user. When Alice or Bob tries to log in to their account, they present their username as Alice/Bob respectively. Here, the username is the identity. Next, the web application will challenge the users to prove their identity (this can be implemented in many ways; for simplicity, we will use unique passwords). This process of verifying the identity of a user is called authentication. After successful login to the applications, each user will be greeted with their personal home screen depending upon their privileges. In our example, Alice, as an admin, will have more capabilities than Bob, as a regular user. These privileges are often stored on the application side (via access control) and are known as authorization. Once logged in, Alice and Bob carry out their duties on the platform. Application often implements rigorous logging systems to track errors, user activity, and potential security incidents. This is called an audit.

This example illustrates the interconnected nature of IAAA. Each element plays a crucial role in ensuring secure access management:

- *Identity* is who you claim you are.

- *Authentication* verifies whether you are who you claim you are.

- *Authorization* outlines what you can do on a system once authenticated.

- *Audit* captures what you did on the system.

By implementing IAAA effectively, organizations can create a robust and secure access management system. This framework, in conjunction with other security controls such as encryption and **data loss prevention** (**DLP**), forms a comprehensive defense against unauthorized access, data breaches, and other cyber threats.

Risk-based approach to security

The digital landscape is a dynamic and ever-evolving terrain, fraught with potential threats and vulnerabilities. Navigating this complex world can feel overwhelming, leaving organizations unsure of where to prioritize their security efforts. Enter the risk-based approach to security, a powerful framework that empowers you to make informed decisions and optimize your security posture.

This section will equip you with essential knowledge and tools to embrace uncertainty and build a security program that truly aligns with your business needs. We will delve into the fundamentals of risk management, equipping you with the ability to identify, analyze, and prioritize potential threats. You will learn about practical risk analysis methodologies and gain hands-on experience through engaging in threat modeling exercises. Ultimately, you will discover the art of balancing risk and business, ensuring that your security efforts deliver tangible value and contribute to your organization's success.

Prepare to shed the burden of uncertainty and embark on a journey toward a more proactive and data-driven approach to security. By the end of this section, you will be empowered to make informed decisions, allocate resources effectively, and build a security program that truly stands the test of time. Let's begin.

Understanding risk management

As defined on Wikipedia (`https://en.wikipedia.org/wiki/Risk_management`) "*Risk management is the identification, evaluation, and prioritization of risks (defined in ISO 31000 as the effect of uncertainty on objectives) followed by coordinated and economical application of resources to minimize, monitor, and control the probability or impact of unfortunate events or to maximize the realization of opportunities.*"

In the ever-changing digital landscape, risk management emerges as the cornerstone of a robust security posture. At its core, risk management involves a systematic process of evaluating potential risks, analyzing their impact and likelihood, and implementing strategies to minimize, mitigate, or transfer these risks. It encompasses a structured approach that enables organizations to anticipate, prepare for, and effectively respond to both internal and external threats that could compromise the integrity, confidentiality, or availability of critical assets.

Before jumping into more detail, let's establish some key nomenclature first:

- An **asset** is any valuable resource that needs protection, including information, systems, applications, and data. This can be data, infrastructure, **intellectual property** (**IP**), trade secrets, or personnel.

- A **threat** represents any potential danger, circumstance, or event capable of causing harm or adversely affecting the confidentiality, integrity, or availability of an organization's assets.

- A **vulnerability** is a weakness or flaw within a system, process, or asset that could be exploited by a threat actor to compromise its security or functionality.

- **Exploitability** is the extent or likelihood to which a vulnerability can be exploited or taken advantage of by an attacker or malicious entity to compromise a system or asset.

Risk is often quantified by the following equation:

$$\text{Risk} = \text{Threat} \times \text{Vulnerability} \times \text{Exploitability}$$

Figure 1.2 – Risk evaluation

The *Risk = Threat × Vulnerability × Exploitability* equation encapsulates the complex interplay between various factors contributing to the overall risk an organization faces. It signifies that risk is not solely determined by the presence of a threat but rather arises from the convergence of multiple elements. A high-impact threat combined with a critical vulnerability and readily available exploitability translates to a significant risk requiring immediate attention. Conversely, a low-impact threat paired with a minor vulnerability and limited exploitability represents a lower risk priority. This formula empowers organizations to prioritize their security efforts effectively and allocate resources strategically to mitigate the most significant risks facing their assets.

Phases

Next, we will go over the risk management process, which typically involves the following key phases:

1. **Identify assets**: The first step involves identifying and classifying all valuable assets within an organization, including information, systems, applications, and data. This comprehensive inventory provides a foundation for understanding the potential impact of security incidents. Often, assigning some dollar values to each identified asset (or potential loss if compromised) becomes crucial in risk-based security models.

2. **Identify threats**: Next, organizations must identify potential threats that could exploit vulnerabilities and compromise their assets. This necessitates ongoing analysis of the evolving threat landscape, considering both external threats such as cyberattacks and internal threats such as human error or malicious insiders.

3. **Assess vulnerabilities**: Vulnerability assessments involve identifying weaknesses within systems, applications, and configurations that could be exploited by identified threats. This analysis helps organizations understand their exposure and prioritize remediation efforts. In addition to conducting vulnerability assessments, evaluating the likelihood of each vulnerability being exploited holds significant importance. This often involves researching readily available exploits on the internet, exposure of systems with vulnerabilities, and so on.

4. **Analyze risks**: By combining the likelihood and impact of identified threats and vulnerabilities, organizations can analyze potential risks associated with each asset. This allows for a data-driven approach to risk prioritization and resource allocation.

5. **Develop risk mitigation strategies**: The core of risk management lies in developing and implementing effective mitigation strategies. This involves a combination of preventative measures such as security controls, detective measures such as monitoring and logging, and recovery strategies to ensure **business continuity (BC)** in the event of an incident.

6. **Continuous monitoring and upgrade**: The risk landscape is constantly evolving, necessitating continuous monitoring and review of the risk management process. This involves updating asset inventories, staying informed of emerging threats and vulnerabilities, and adjusting mitigation strategies as needed. This is one of the most vital components of any security program; an outdated risk mitigation strategy is sometimes worse than having no plans in place.

Advantages

Implementing a robust risk management framework offers numerous benefits for organizations:

- **Improved decision-making**: Risk analysis provides data-driven insights for prioritizing security investments and allocating resources effectively.

- **Proactive approach**: By identifying and mitigating risks proactively, organizations can prevent incidents before they occur, minimizing damage and downtime.

- **Enhanced security posture**: Risk management fosters a culture of security awareness and drives continuous improvement of security controls and processes.

- **Compliance**: Many regulations require organizations to implement risk management frameworks, ensuring compliance and avoiding penalties.

Understanding risk management provides a solid foundation for building a secure and resilient digital environment. By systematically identifying, analyzing, and mitigating potential threats, organizations can proactively protect their valuable assets and achieve their business goals.

Risk analysis

Now that we have established a solid foundation of the need for risk management, in this section, we will focus on one core aspect of it: risk analysis. Risk analysis is a multifaceted process crucial for evaluating and understanding potential threats and vulnerabilities within an organizational framework. It involves a systematic examination of risks to assess their likelihood of occurrence, potential impact, and the effectiveness of existing controls. This process aims to identify, prioritize, and manage risks that could compromise an organization's objectives, assets, or operations. It delves deeper than mere identification and assessment, providing a comprehensive understanding of potential risks and their implications for an organization. One common tool for that is the risk matrix. Depending on the likelihood of a vulnerability getting exploited and impact of such an event; we can estimate potential risk using such matrix.

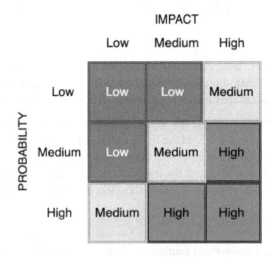

Figure 1.3 – 3x3 risk matrix

Organizations commonly adopt two primary approaches to analyzing risks: *qualitative* and *quantitative* analysis. On one hand, qualitative analysis involves subjective assessments, typically utilizing descriptive scales or categories to evaluate risks based on their severity, likelihood, and impact. In contrast, quantitative analysis involves a more data-driven approach, leveraging numerical data and statistical models to measure and quantify risks.

Qualitative analysis

Qualitative analysis utilizes descriptive methods to assess the likelihood and impact of potential threats and vulnerabilities. This approach employs expert judgment, risk matrices, and brainstorming sessions to identify and prioritize risks based on factors such as industry best practices, historical incidents, and regulatory requirements. The benefit of qualitative analysis lies in its versatility and adaptability, allowing organizations to analyze a wide range of risks and scenarios, regardless of readily available data:

Quantitative analysis

Quantitative analysis leverages data and statistics to estimate the likelihood of threats and the potential financial impact of their occurrence. This approach involves techniques such as attack trees, fault trees, and Monte Carlo simulations to generate numerical estimates of risk. The advantage of quantitative analysis lies in its objectivity and data-driven approach, providing a more precise assessment of risk severity and facilitating informed resource allocation decisions.

Effective risk analysis often involves a judicious blend of both qualitative and quantitative methodologies. By leveraging the strengths of each approach, organizations can gain a holistic understanding of their risk landscape and make well-informed decisions about their security posture. While qualitative analysis can help identify and prioritize emerging risks, offering insights into the nature and context of risks, quantitative analysis can provide precise information that aids in making data-driven decisions to support resource allocation and prioritizing risk management efforts.

This holistic approach enables organizations to effectively manage risks by considering both qualitative aspects and quantitative data, ensuring a more robust and well-informed risk management strategy.

Threat modeling

As we covered the risk-based security model for organizations, the identification of threats and assets is a key step in the process. In this section, we will go over the concept of threat models and how organizations can effectively employ them to build a robust security posture.

Threat modeling is a proactive approach to security that anticipates and mitigates potential threats before they materialize. By systematically analyzing systems and applications from an attacker's perspective, it identifies vulnerabilities and recommends safeguards to prevent successful exploits. This proactive approach delivers numerous benefits, including the following:

- **Reduced security costs**: By focusing resources on mitigating high-impact risks, organizations can optimize their security investments.

- **Improved security posture**: Identifying and addressing vulnerabilities early on reduces the likelihood of successful attacks.

- **Enhanced decision-making**: Threat modeling provides data-driven insights to support informed decisions about security controls and resource allocation.

- **Improved communication and collaboration**: Threat modeling fosters a shared understanding of security risks and facilitates collaboration between security and development teams.

Understanding threat modeling frameworks

Securing digital assets in the face of ever-evolving threats demands a comprehensive understanding of potential vulnerabilities and attack strategies. Fortunately, various frameworks offer invaluable guidance in analyzing and mitigating security risks.

Several frameworks exist for conducting threat modeling, each with its own methodologies and focuses. The **Spoofing, Tampering, Repudiation, Information disclosure, DoS, Elevation of privilege (STRIDE)** and **Damage, Reproducibility, Exploitability, Affected users, Discoverability (DREAD)** models are widely employed frameworks. STRIDE helps identify different types of threats, while DREAD provides a scoring mechanism to prioritize threats based on their severity. Additionally, the **Process for Attack Simulation and Threat Analysis (PASTA)** methodology is highly regarded for its risk-centric approach, focusing on the attacker's motivations and objectives.

Let's briefly cover some of these very common threat modeling frameworks as they can provide a structural process:

- **MITRE ATT&CK:**

    - Identifies and catalogs adversary TTPs across diverse platforms and operating systems.

    - Enables proactive identification and mitigation of potential threats based on real-world adversary behavior.

    - Provides a structured knowledge base for conducting threat modeling and adversary emulation exercises.

- **STRIDE:**

    - Focuses on six major threat categories: Spoofing, Tampering, Repudiation, Information Disclosure, DoS, and Elevation of privilege.

    - Enables identification and prioritization of vulnerabilities based on potential impact and ease of exploitation.

    - Offers a simple and intuitive framework that is easily understandable by non-technical stakeholders.

- **PASTA:**

    - Emphasizes the identification of attack trees, which represent potential attack scenarios and their required conditions.

    - Facilitates a structured analysis of attack paths and helps identify critical vulnerabilities that could lead to compromise.

    - Provides a visual representation of potential threats, enhancing communication and collaboration among security teams.

- **NIST Cybersecurity Framework (CSF):**

    - Outlines five core functions: **Identify, Protect, Detect, Respond, and Recover (IPDRR)**

    - Offers a comprehensive roadmap for developing and implementing effective cybersecurity programs.

- Provides a flexible and adaptable framework that can be tailored to diverse organizational needs and risk profiles.

- **Operationally Critical Threat, Asset, and Vulnerability Evaluation (OCTAVE):**

    - Focuses on operational risk management, helping organizations prioritize and address high-impact threats in critical systems.

    - Utilizes a risk-driven approach to identify and assess vulnerabilities based on potential impact and likelihood of occurrence.

    - Facilitates collaboration between security and operational teams, ensuring that risk management efforts are aligned with operational priorities.

- **TRIKE:**

    - Analyzes security threats from an attacker's perspective, focusing on their goals, resources, and initial access points.

    - Provides a structured approach for identifying attack paths and prioritizing vulnerabilities based on their exploitability.

    - Offers a valuable tool for conducting threat modeling exercises and developing comprehensive risk mitigation strategies.

Choosing the right framework

Each framework offers unique benefits and caters to specific needs:

- **MITRE ATT&CK**: Ideal for organizations facing complex and sophisticated threats, requiring a deep understanding of adversary TTPs.

- **STRIDE**: Suitable for identifying and prioritizing vulnerabilities in software applications and web services.

- **PASTA**: Beneficial for visually representing potential attack scenarios and facilitating collaborative risk analysis.

- **NIST CSF**: Provides a comprehensive roadmap for developing and implementing a holistic cybersecurity program.

- **OCTAVE**: Effective for managing operational risks and prioritizing critical security investments.

- **TRIKE**: Useful for conducting threat modeling exercises and analyzing attack paths from an attacker's perspective.

The choice of framework depends on several factors, including the following:

- **Organizational needs and risk profile**: Different organizations face varying threats and have unique security requirements.

- **Technical expertise and resources**: Some frameworks require more technical expertise than others.

- **Desired outcomes**: Different frameworks cater to different goals, such as identifying vulnerabilities, prioritizing risks, or developing mitigation strategies.

No single framework provides a one-size-fits-all solution to security risk management. By understanding the strengths and limitations of each framework, organizations can leverage a multi-framework approach to gain a comprehensive understanding of their threat landscape and develop effective mitigation strategies. This ensures that their valuable assets are protected in the face of evolving threats.

Hands-on threat modeling

Threat modeling often starts with developers bringing their ideas to a security team. In this section, let's embark on a threat modeling journey, dissecting the security of a fictional banking application, **FinBank Mobile**. We'll use the STRIDE framework to identify potential threats and vulnerabilities:

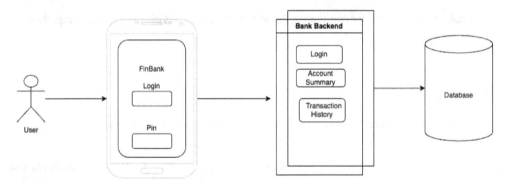

Figure 1.4 – Overview of FinBank Mobile application

The preceding diagram highlights the mobile application and a very high-level architecture of the design. Here are a few important properties:

- FinBank Mobile is a mobile application that allows users to access their accounts, view transactions, transfer funds, and make payments.

- Users authenticate through a PIN or fingerprint scanner.

- The app connects to FinBank's backend servers via secure communication protocols.

The very first step of any threat model is to identify critical assets, such as the following:

- User credentials (logins, PINs)

- Account information (balances, transaction history)

- Financial data (funds, payment details)

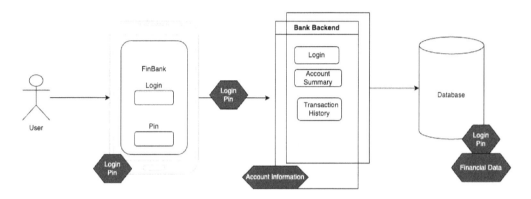

Figure 1.5 – Critical assets for FinBank Mobile application threat model

Next, let's look at a typical data flow pattern often known as a **data flow diagram** (DFD):

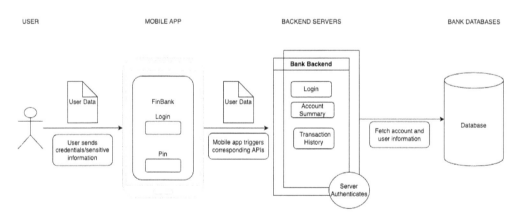

Figure 1.6 – DFD for FinBank

Data flow often aids security professionals in better understanding the expected pattern of information flow within the system. It is very important to understand the use case to anticipate what might go wrong. Now, let's explore some threats by following the STRIDE framework and exploring some mitigation strategies.

Spoofing

- **Threat**: Attacker impersonates a legitimate user to access accounts
- **Vulnerability**: Weak authentication mechanisms, lack of **two-factor authentication (2FA)**
- **Mitigation**: Strong passwords, MFA, device verification

Tampering

- **Threat**: Attacker modifies transaction data or account information
- **Vulnerability**: Unencrypted data transmission, insecure backend systems
- **Mitigation**: Data encryption in transit and at rest, secure coding practices, server hardening

Repudiation

- **Threat**: Attacker denies making a transaction or claims unauthorized access
- **Vulnerability**: Lack of transaction logging, weak audit trails
- **Mitigation**: Secure logging of all user actions and transactions, tamper-proof audit trails

Information disclosure

- **Threat**: Attacker steals sensitive user data or account information
- **Vulnerability**: Insecure APIs, data breaches, malware on user devices
- **Mitigation**: Secure API design, robust data encryption, user education on malware threats

DoS

- **Threat**: Attacker disrupts app functionality or server availability
- **Vulnerability**: Unsecured network infrastructure, poor coding practices
- **Mitigation**: **Distributed DoS (DDoS)** protection, secure coding practices, server redundancy

Elevation of privilege

- **Threat**: Attacker gains unauthorized access to other users' accounts or system administrator privileges
- **Vulnerability**: Weak access controls, insufficient user authorization
- **Mitigation**: **Principle of least privilege (PoLP)**, robust access control mechanisms, regular security audits

Now that we have identified some threats, it is important to rank them based on their severity, likelihood, and potential impact. Remember—threat modeling is an iterative process. This example is a simplified illustration, and real-world scenarios might involve more complex threats and vulnerabilities. However, it serves as a valuable starting point for understanding and mitigating risks in your own systems.

By proactively identifying these vulnerabilities, the development team can implement appropriate security controls, such as MFA, data encryption, and secure communication protocols. This proactive approach significantly reduces the risk of successful attacks and protects the confidentiality, integrity, and availability of sensitive financial data.

Several tools and resources can aid in threat modeling exercises, such as the following:

- **Threat Dragon**: A web-based application that facilitates collaborative threat modeling.

- **Attack trees**: A browser extension that helps visually represent attack scenarios.

- **IriusRisk**: A comprehensive threat modeling platform with advanced features and integrations.

By leveraging threat modeling methodologies and available tools, organizations can gain a clear understanding of their potential security risks and implement effective mitigation strategies to protect their valuable assets.

Balancing risk with business needs

As we conclude this comprehensive exploration of risk-based security practices, it becomes imperative to underscore the importance of balancing risk management with the strategic needs of the business. Implementing a robust risk-based security approach ultimately boils down to achieving a delicate balance between security and business needs. While comprehensive mitigation strategies are crucial, over-securing can stifle innovation and hinder business growth. Risk analysis plays a pivotal role in this balancing act, providing quantifiable data to inform decision-making. Just as it wouldn't be prudent to deploy a million-dollar security control for a hundred-dollar asset, prioritizing resources effectively becomes paramount.

Remember—absolute security is an elusive ideal. No system is truly 100% secure, as attackers constantly evolve their tactics and exploit emerging vulnerabilities. Instead of chasing an unattainable objective, the focus should shift toward building a resilient and adaptable security posture. This means prioritizing high-impact risks while accepting and managing residual risks through a combination of controls, detection mechanisms, and **incident response** (**IR**) capabilities.

By embracing a risk-based approach and fostering a collaborative environment where business and security teams work together, organizations can achieve a sustainable and effective security posture. This approach allows them to focus their efforts on the most critical threats, optimize resource allocation, and ultimately achieve their business goals while ensuring their valuable assets remain secure.

Identifying threat actors and understanding their motivations

Building upon our exploration of risk-based security, threat modeling, and risk analysis, the upcoming section embarks on a journey to delve deeper into the realm of those who seek to exploit these vulnerabilities: **threat actors**. In the realm of cybersecurity, comprehending the intricate motivations and behaviors of threat actors stands as a pivotal piece in fortifying an organization's defenses. Understanding their motivations, intentions, and strategies provides a nuanced perspective crucial for anticipating, mitigating, and effectively countering potential cyber threats. By unveiling the intricacies of threat actors' mindsets and objectives, we equip ourselves with insights vital for crafting resilient security strategies and proactive defenses against an evolving array of cyber risks.

This section will expose the inner workings of these adversaries, unveil their diverse motivations, and explore the spectrum of tactics they employ. Armed with this knowledge, we can bolster our defenses, minimize potential damage, and ensure the resilience of our digital assets. Prepare to embark on a journey into the minds of the adversaries, where understanding is the ultimate weapon in the battle for security.

Types of attackers

In the vast realm of cybersecurity, threat actors come in diverse forms, each with their unique motives and approaches. Imagine stepping into this digital world—there are the script kiddies, the rookies of this realm, often dabbling with existing tools and techniques out of curiosity or a quest for recognition. They're akin to fledgling explorers, testing the waters, yet not fully comprehending the depth of their actions.

Moving further along, you encounter hackers, individuals equipped with more skills and knowledge. Their motivations vary—some seek financial gains; others pursue ideological agendas. They're like crafty adventurers, using their expertise to navigate through digital systems, aiming for treasures or unlocking gates for their beliefs.

Yet, as you venture deeper, you stumble upon organized cybercrime syndicates, akin to well-established guilds in this digital landscape. These groups operate like sophisticated businesses, meticulously planning and executing attacks on organizations. Their motives? Often financial—seeking profits through ransomware, data theft, or other nefarious activities.

As you journey further into this digital landscape, you encounter state-sponsored attackers—operatives backed by governments or nation-states. Their quests are grander, driven by political, economic, or military goals. They conduct elaborate schemes—digital espionage, cyber warfare, or strategic intelligence gathering—impacting not just organizations but entire nations.

But watch your steps closely, for within these digital realms, even those you trust could pose a threat. Insider threats, whether through negligence, compromised credentials, or intentional malice, originate from within organizations. They're like shadows—sometimes overlooked, yet capable of causing significant harm.

Understanding these diverse personas within the cyber realm allows organizations to craft tailored defenses. From fortifying basic defenses against curious explorers to building intricate shields against the grand schemes of nation-states, the landscape of cybersecurity is as dynamic and multifaceted as the threats it seeks to counter.

Now, let's look at them from another lens: motivations.

Threat actor motivations

The digital realm attracts a diverse array of adversaries, each driven by unique motives and wielding distinct capabilities. Understanding these different types of attackers is crucial for tailoring our security posture to effectively counter their threats:

- **Financially motivated**: This ubiquitous category encompasses individuals and groups driven by the pursuit of financial gain. They may launch cyberattacks to steal sensitive financial information, extort funds through ransomware campaigns, or disrupt critical infrastructure for financial benefit. Cybercrime syndicates, financially motivated hacktivists, and lone wolves seeking personal gain all fall under this umbrella.

- **Ideologically driven**: Fueled by political, religious, or social beliefs, these attackers aim to disrupt operations, spread propaganda, or cause damage aligned with their ideology. They may launch DoS attacks, deface websites, or leak sensitive information to achieve their objectives. Hacktivist groups, nation-states engaging in cyber warfare, and individuals driven by extremist ideologies fall within this category.

- **Espionage and cyber warfare**: Governments and intelligence agencies engage in cyber espionage to steal classified information, conduct surveillance, and gain advantage in geopolitical conflicts. These attacks are often highly sophisticated and targeted, employing advanced techniques and zero-day exploits. Nation-states and their sponsored actors are primary actors in this category.

- **Insider threats**: Often overlooked but potentially devastating, insider threats arise from individuals with authorized access who misuse their privileges for malicious purposes. This could include disgruntled employees, contractors with financial motives, or individuals coerced by external actors. Detecting and mitigating insider threats requires a combination of technical controls, awareness training, and robust access control policies.

- **Script kiddies and advanced persistent threats (APTs)**: The landscape also includes less skilled individuals known as "script kiddies" who exploit readily available tools and scripts to launch basic attacks. Conversely, APTs represent highly skilled and resourced groups capable of conducting sophisticated and sustained attacks. Understanding the capabilities and motivations of both extremes is crucial for developing comprehensive defenses.

By recognizing the diverse motivations and capabilities of these threat actors, organizations need to tailor their security strategies to effectively counter their tactics and protect their valuable assets.

Real-world examples

The digital realm is a playground for adversaries, with motivations and skill levels as diverse as the tools at their disposal. On one end of the spectrum, we have script kiddies, teenage hackers who hop on the latest vulnerability bandwagon, wielding readily available exploits found on forums and exploit databases. Imagine a teenager stumbling upon a SQL injection exploit in a popular online forum, then excitedly testing it on various websites for the thrill of seeing a server cough up data. While their attacks may be unsophisticated and often ineffective, they serve as a constant reminder of the importance of patching vulnerabilities promptly.

Moving further along the spectrum, we encounter APTs, well-organized groups with a level of skill and resources that rivals some nation-states. Take **The Shadow Brokers** (**TSB**), a mysterious hacking group that surfaced in 2016, leaking a trove of **National Security Agency** (**NSA**) hacking tools. Their exploits, dubbed EternalBlue and DoublePulsar, targeted vulnerabilities in Windows software, giving attackers unprecedented access to millions of computers worldwide. In 2017, WannaCry, a ransomware attack powered by EternalBlue, crippled hospitals, banks, and government agencies across the globe. This attack served as a stark reminder of the devastating impact APTs can have when they unleash their potent arsenal.

The contrast between script kiddies and TSB highlights the dynamic nature of the threat landscape. While script kiddies may rely on readily available tools and opportunistic exploits, APTs invest heavily in research and development, crafting sophisticated attack vectors and exploiting vulnerabilities before vendors even know they exist. This evolution necessitates a multilayered approach to security, encompassing robust patching strategies, advanced TI, and proactive network monitoring to detect and thwart attacks – regardless of their source.

By understanding the motivations and capabilities of these diverse adversaries, organizations can build a more effective security posture—one that not only shields against the script kiddie with a copied exploit but also stands resilient against the sophisticated tactics of an APT such as TSB.

Security through the ages

Security, the ever-vigilant guardian of our digital realm, has evolved as dramatically as the technology it protects. Like a chameleon adapting to its environment, the focus of security researchers and practitioners has shifted through the ages, reflecting the changing threats and vulnerabilities of each era. This section takes you on a captivating journey through the history of security, unveiling how different times have shaped the priorities and practices of this critical discipline.

From the early days of mainframes, where physical access and data breaches were the primary concerns, to the internet-fueled era of cyberattacks and malware, security has constantly adapted to meet the challenges of a rapidly evolving landscape. We'll explore the emergence of landmark frameworks such as the **Open Worldwide Application Security Project (OWASP)** Top 10, which standardized security best practices and empowered developers to build secure systems. We'll delve into the transformative impact of cloud computing, where the shared responsibility model redefined the security landscape and necessitated a collaborative approach.

Ultimately, this section will not only illuminate the fascinating history of security but also offer valuable insights for the present. You'll gain a deep understanding of how security priorities have shifted, what challenges lie ahead, and why security has become a non-negotiable top priority for every organization in the digital age. So, buckle up, curious readers, as we embark on this captivating exploration of security's timeless dance with ever-evolving threats.

Trends in security

Imagine security as a chameleon, constantly adapting its colors to blend in with the ever-shifting terrain of digital threats. In the mainframe era, it focused on physical access control, guarding fortresses of data with the diligence of a medieval knight. As the internet emerged, it donned the cloak of a vigilant sentry, scanning for script kiddies wielding basic tools such as dial-up modems. The Wild West of early web development saw the rise of OWASP Top 10, a beacon of best practices in the face of rampant SQL injection and **cross-site scripting (XSS)** attacks.

The mobile revolution brought a new set of challenges, demanding secure coding practices and data encryption like a digital alchemist crafting impenetrable shields. Cloud computing, with its shared responsibility model, blurred the lines of defense, necessitating a collaborative approach. APIs, the lifeblood of modern applications, became vulnerable targets, prompting OWASP Top 10 to evolve in response, urging developers to bake security into their code from the ground up.

This shift toward "shifting left" marked a pivotal moment. Security was no longer a reactive afterthought stationed at the end of the development pipeline. Instead, it became an active participant, guiding developers with OWASP Top 10 as a shared playbook. Static code analysis tools, akin to X-ray glasses for code, emerged to identify vulnerabilities before deployment, while automation and **continuous integration/continuous deployment (CI/CD)** pipelines seamlessly integrated security checks into the development workflow.

Today, the landscape is a dynamic ecosystem of microservices, containers, and serverless functions. OWASP Top 10 reflects this complexity, highlighting supply chain vulnerabilities and broken access control like a seasoned cartographer navigating uncharted territory. DevSecOps, a fusion of development, security, and operations, has become the mantra, ensuring that security is not merely an add-on but an intrinsic part of the development process.

Security's metamorphosis is not just a trend but a strategic imperative. In the ever-evolving digital jungle, only those who build security into their DNA can truly thrive. So, join us as we delve deeper into the world of cloud computing and explore how it is transforming the security landscape, one line of code at a time.

The rise of cloud computing

The ascension of cloud computing ushered in a tectonic shift in how organizations perceive, adopt, and implement security measures. This technology revolutionized the very fabric of traditional IT infrastructures, offering scalability, flexibility, and cost efficiency previously unattainable through on-premises systems. However, its emergence wasn't just a transformation in computational paradigms; it was a seismic recalibration of security models for businesses worldwide.

The arrival of cloud computing wasn't just a technological advancement; it was a seismic shift in the security landscape. Suddenly, the traditional fortress-like model of on-premises data centers crumbled, replaced by a shared responsibility model where security wasn't just an IT department concern but a collaborative effort between providers and users. This paradigm shift fundamentally changed how everyone, from individuals to enterprises, approached securing their digital assets.

Firstly, cloud providers such as **Amazon Web Services (AWS)**, Microsoft Azure, and **Google Cloud Platform (GCP)** offered robust, scalable security infrastructure. They invested in advanced data centers, IDSs, and TI, providing a level of security that many organizations could never afford on their own. This democratized access to high-grade security, allowing even small businesses to benefit from the expertise and resources of industry giants.

However, the shared responsibility model also introduced new challenges. While cloud providers secured the infrastructure, organizations remained responsible for securing their data and applications within the cloud environment. This meant implementing proper access controls, encryption protocols, and vulnerability management practices. The rise of cloud-specific vulnerabilities, such as insecure API integrations and misconfigurations, demanded new skills and awareness from users, blurring the lines between traditional IT security and cloud security expertise.

Ultimately, cloud computing redefined the security equation, forcing organizations to embrace a collaborative approach. It fostered communication and transparency between providers and users, encouraging joint efforts to identify and mitigate threats. The rise of security frameworks such as the **Cloud Security Alliance's (CSA's) Cloud Controls Matrix (CCM)** provided a common language and best practices for navigating this new shared landscape.

Moreover, the move to the cloud amplified the need for a holistic and adaptable security posture. Organizations embraced a more dynamic approach, recognizing that security must be agile and elastic, capable of swiftly adapting to the dynamic and expansive nature of cloud ecosystems. This paradigm shift, although disruptive, empowered businesses to leverage advanced security tools and frameworks offered by cloud providers, fostering a more robust security culture across industries. The rise of cloud computing etched a new normal – a landscape where security became an integral, inseparable facet of the cloud fabric, forever altering the way organizations conceptualize and implement security strategies.

Security is omnipresent

Security, once a niche concern confined to IT departments and government agencies, has undergone a dramatic transformation. In today's hyper-connected world, it's no longer a luxury but a ubiquitous necessity that permeates every facet of our digital lives. From the moment we wake up and check our smartphones to the seamless transactions we make online, security forms the invisible backbone of our digital experiences.

This omnipresence stems from the fundamental shift in how we interact with technology. Our personal and professional lives are increasingly intertwined with online platforms, storing sensitive data and facilitating critical transactions. This inherent dependence on digital infrastructure creates a vast attack surface for malicious actors, making robust security a non-negotiable priority for individuals and organizations alike.

The rise of cybercrime, with its ever-evolving tactics and sophisticated tools, has further fueled this sense of urgency. Data breaches, ransomware attacks, and identity theft are no longer distant threats but tangible realities that can impact anyone, from small businesses to global corporations. This has propelled security to the forefront of boardroom discussions and individual decision-making, driving investments in security technologies and fostering a culture of security awareness.

However, security's ubiquity goes beyond mere protection. It has become an essential enabler of trust and confidence in the digital ecosystem. Secure online transactions fuel e-commerce, secure communication platforms empower remote work, and secure data storage unlocks the potential of cloud computing. In essence, security is the foundation upon which the digital world thrives, fostering innovation and collaboration while mitigating risk and ensuring the integrity of our digital interactions.

This ubiquitous nature of security demands a multi-pronged approach. Individuals must adopt safe online practices, maintain strong passwords, and be cautious about their digital footprint. Organizations need to invest in robust security infrastructure, implement comprehensive security policies, and prioritize employee training. Governments have a crucial role in fostering collaboration among stakeholders, developing regulations, and promoting cybersecurity awareness.

In this landscape, where the digital permeates every aspect of existence, security's elevation to a top priority is no longer a matter of debate but an incontrovertible truth. Ultimately, security is not simply a technical challenge but a collective responsibility. By recognizing its omnipresence and prioritizing its implementation, we can create a safer, more trustworthy, and more resilient digital landscape for all.

Summary

In today's digital epoch, security has transcended its role as a mere addendum to becoming the linchpin of every facet of our interconnected world. While security has matured over the past few decades and become more complex, fundamental aspects of the subject are still intact. In this chapter, we took a ride through time and developed intuitions about the emergence of security we experience today. We hope this chapter helped bolster your security thought process, as we will be coming back to a lot of these concepts throughout the book.

Key takeaways

- The CIA Triad: confidentiality, integrity, and availability—the three core pillars of security. All security incidents can be attributed to the violation of one or more of these principles.

- Security is not just about identifying gaps and deploying assets. Security involves technical controls such as encryption, authorization, and so on, as well as procedural controls such as adhering to standards, policies, and guidelines.

- In the realm of security, risk is often calculated from the confluence of threat, vulnerability, and exposure. Understanding risk in security requires a holistic perspective, considering not just potential threats and system vulnerabilities but also the context of exposure and potential consequences.

- Identifying all critical assets for your organization often marks major progress toward building a resilient security posture.

- Threat modeling not only helps organizations understand the pertinent threat landscape they face but instills a deep security mindset within the team.

- In the world of security, change is the only constant. Organizations need to adjust their defense posture with the evolution of new attack vectors.

Building on top of risk-based security, next, we will delve into the concept of DiD and how you can build a security program that can withstand the ever-changing landscape of security.

Congratulations on getting this far; give yourself a pat on the back – you deserve it!

Further reading

To learn more about the topics that were covered in this chapter, take a look at the following resources:

- Charles Babbage: `https://www.computerhistory.org/babbage/`

- The Creeper virus: `https://www.historyofinformation.com/detail.php?entryid=2860`

- ARPANET: `https://www.darpa.mil/about-us/timeline/arpanet`

- LaPadula model: `https://www.sciencedirect.com/topics/computer-science/lapadula-model`

- How Alan Turing Cracked The Enigma Code: `https://www.iwm.org.uk/history/how-alan-turing-cracked-the-enigma-code`

- The Morris Worm: `https://www.fbi.gov/news/stories/morris-worm-30-years-since-first-major-attack-on-internet-110218`

- The Melissa Virus: `https://www.fbi.gov/news/stories/melissa-virus-20th-anniversary-032519`

- Titan Rain: `https://www.cfr.org/cyber-operations/titan-rain`

- WannaCry: `https://www.cisa.gov/sites/default/files/FactSheets/NCCIC%20ICS_FactSheet_WannaCry_Ransomware_S508C.pdf`

- The untold story of SolarWinds: `https://www.npr.org/2021/04/16/985439655/a-worst-nightmare-cyberattack-the-untold-story-of-the-solarwinds-hack`

- ISO 31000 Risk management: `https://www.iso.org/iso-31000-risk-management.html`

- MITRE ATT&CK framework: `https://attack.mitre.org/`

- Threat modeling: `https://owasp.org/www-community/Threat_Modeling_Process`

- OWASP Threat Dragon: `https://owasp.org/www-project-threat-dragon/`

- IriusRisk: `https://www.iriusrisk.com`

- Edward Snowden, the NSA whistleblower: `https://www.theguardian.com/world/2013/jun/09/edward-snowden-nsa-whistleblower-surveillance`

- The Shadow Brokers: `https://en.wikipedia.org/wiki/The_Shadow_Brokers`

- **Cybersecurity and Infrastructure Agency (CISA)** insider threats: `https://www.cisa.gov/topics/physical-security/insider-threat-mitigation/defining-insider-threats`

- APTs: `https://www.cisa.gov/topics/cyber-threats-and-advisories/advanced-persistent-threats-and-nation-state-actors`

- OWASP Top 10: `https://owasp.org/www-project-top-ten/`

- Shift left security: `https://www.crowdstrike.com/cybersecurity-101/shift-left-security/`

- Shared Responsibility Model: `https://aws.amazon.com/compliance/shared-responsibility-model/`

- CSA CCM model: `https://cloudsecurityalliance.org/research/cloud-controls-matrix/`

- Importance and relevance of CIA Triad: `https://www.cisecurity.org/insights/spotlight/ei-isac-cybersecurity-spotlight-cia-triad`

2

Practical Guide to Defense in Depth

In the treacherous landscape of cybersecurity, a single vulnerability can be your undoing. This is the stark reality captured in the adage: *"Security is a chain, and the weakest link breaks it."* Traditional risk assessments and threat modeling identify critical gaps within our systems, but they often fail to account for the inherent fragility of a single-layered defense.

Imagine your digital assets as a prized castle. A determined adversary needs only one breach to plunder its treasures. **Defense in depth** (**DiD**), a well-recognized yet underutilized strategy, flips the script. By strategically placing multiple layers of security controls around your assets, you force attackers to navigate a labyrinth of obstacles. Like a sturdy castle wall, each layer increases the cost and complexity of intrusion, deterring even the most persistent adversaries.

Understanding the core principles and practical application of DiD is no longer a luxury but a necessity. This chapter delves into the heart of this multifaceted strategy. We will dissect the security landscape into distinct domains, explore the value proposition of each control layer, and equip you with the knowledge to craft a robust, multilayered security framework. By the end of this journey, you will possess the tools and insights to transform your digital fortress into an impenetrable bastion.

In this chapter, we're going to cover the following main topics:

- The concept of DiD
- Security domains and controls
- Selecting and implementing the right controls
- Glimpse of a real-world DiD approach

Let's get started!

The concept of DiD

In the perilous landscape of cybersecurity, a single vulnerability can be the chink in your armor, rendering even the most robust defenses ineffective. Just imagine your prized data locked away in a high-security vault, protected by intricate locks and sophisticated alarm systems. Yet, amid these safeguards, a seemingly innocuous backdoor – a forgotten service port left open – grants an opportunistic hacker easy access, turning your fortress into a playground.

This scenario isn't mere conjecture. Recent attack simulations by cybersecurity researchers highlight the alarming reality of single-point failures. In October 2023, a simulated cyberattack on a major hospital exposed a critical vulnerability in its medical device communication protocols. By exploiting a single unpatched software flaw, attackers gained access to sensitive patient data, demonstrating the devastating consequences of neglecting even the smallest security gap.

DiD emerges as the vital antidote to such vulnerabilities. This multilayered approach to security mimics the layered defenses of a medieval castle, with each layer acting as a formidable obstacle for attackers to overcome. Imagine your vault now shrouded within concentric rings of security: firewalls acting as the outer moat, **intrusion detection systems (IDSs)** as vigilant guards, and data encryption as an additional, unbreakable lock. Each layer buys precious time, increases the attacker's effort, and ultimately raises the cost of a successful breach.

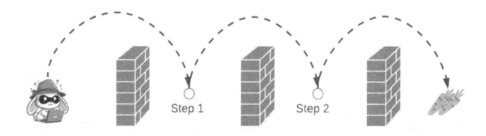

Figure 2.1 – Demonstration of multilayered security posture

By embracing DiD, you transform your security posture from a fragile shell to an impenetrable fortress. For the rest of this chapter, we will delve into the practical application of this multifaceted strategy, equipping you with the knowledge and tools to build a resilient defense against ever-evolving cyber threats.

The fallacy of single-point defense

Imagine a chain holding back a precious treasure. Forged from the toughest steel, its links appear impenetrable. Yet, what if a single rusty link buckles under pressure? The entire chain fails, and the treasure is lost. This is the harsh reality of single-layered security: seemingly robust but fundamentally vulnerable to even the smallest chink in its armor.

In the digital realm, this fragile chain manifests in various forms. Consider a company relying solely on a firewall to protect its network. Like a lone gatekeeper, the firewall stands guard against unauthorized access. However, a single unpatched software vulnerability or a cleverly crafted phishing attack can bypass the firewall, granting access to the entire network and its sensitive data. Similarly, relying solely on antivirus software, while invaluable, leaves your system exposed to zero-day exploits or attacks targeting specific applications.

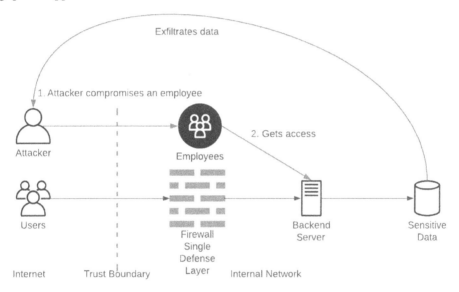

Figure 2.2 – Fallacy of Single-Point Defense

The fallacy of single-layered security hinges on its limited scope. It focuses on defending a single point, neglecting the vast attack surface surrounding it. Think of the attack surface as the total area exposed to potential intruders. A single-layered defense only covers a fraction of this surface, leaving countless entry points for attackers to exploit.

A crucial aspect accentuating the vulnerability of a singular security layer is the concept of the "attack surface" [1]. This refers to the total points, both tangible and digital, through which an attacker can infiltrate or compromise a system. In a single-layered security paradigm, the attack surface remains vast, offering attackers a wide array of potential entry points.

A real-world anecdote [2]

In the 2020 Twitter hack, several high-profile Twitter accounts, including those of prominent individuals and organizations such as Elon Musk, Barack Obama, and Apple, were compromised in a coordinated attack. The attackers managed to gain access to the accounts and posted tweets soliciting Bitcoin transfers, promising to double the amount sent. The fraudulent tweets aimed to deceive followers into sending cryptocurrency to the attackers' wallets.

The attack was carried out through a social engineering scheme targeting Twitter employees, gaining access to internal systems and tools. This breach allowed the attackers to reset email addresses associated with the accounts and bypass **two-factor authentication** (**2FA**), granting them control over the compromised profiles.

The incident raised concerns regarding the security of high-profile social media accounts, highlighting the vulnerabilities within centralized systems and emphasizing the importance of robust security measures, especially in platforms with global reach and significant user bases.

This attack highlights how a weakness in a computer system can be exploited by attackers to gain an initial foothold. Thankfully, there's a more robust approach: DiD. This strategy builds upon the principle of redundancy, adding multiple layers of security around your critical assets. Each layer acts as a barrier, forcing attackers to navigate an intricate maze of obstacles before reaching the core. Firewalls, IDSs, data encryption, and application security controls – each layer shrinks the attack surface, exponentially increasing the attacker's effort and cost of success.

Understanding the limitations of single-layered security is the first step to transforming your digital realm from a vulnerable chain to an impenetrable citadel.

Diversification of defense

In the ongoing battle against cyber threats, the "strength in numbers" ground truth finds new relevance through the strategic deployment of diverse security controls. Rather than relying on a single impervious shield, organizations are bolstering their defenses by employing a tapestry of varied security measures, each addressing specific attack vectors. This mosaic of defenses serves as an effective deterrent against a myriad of common attack techniques that assail modern systems.

Imagine a castle guarded by a single, stagnant moat. While initially formidable, a clever attacker with enough patience and the right tools could eventually breach it. This is precisely the dilemma faced by security architects relying solely on a single defense mechanism. While layered security offers a sturdy outer wall, relying solely on one type of brick within that wall invites exploitation by attackers constantly honing their tools and tactics. Diversification of defenses is a critical principle in DiD that strengthens resilience by embracing heterogeneity and adaptability.

Gone are the days when a singular firewall or antivirus could safeguard your domain. Today's attackers wield intricate arsenals, evolving their techniques alongside defensive advancements. A static, homogenous defense becomes a predictable canvas for their artistry. Diversification, on the other hand, throws a chaotic wrench into their plans. Think of it as constructing your castle with walls of varied height and material, interspersed with hidden traps and vigilant guards. Each layer presents a unique challenge, forcing attackers to expend resources and time navigating a labyrinth of complexity. From firewalls and IDSs to **multi-factor authentication** (**MFA**) and encryption protocols, these security controls create a formidable barrier, each layer fortified to counter specific attack methods.

The benefits of embracing diverse defenses are multifold. Redundancy ensures that even if one layer falls, the overall security fabric remains strong. Failover mechanisms seamlessly redirect compromised traffic, minimizing downtime and potential damage. Deception, such as honeytraps and phantom servers, wastes attacker resources and reveals their tactics. Each diverse control, be it preventive, detective, or corrective, adds a valuable piece to the security puzzle, hindering attackers at every turn.

Imagine a scenario where an attacker launches a phishing campaign. Your diversified defense kicks in:

- Email gateways with diverse detection algorithms filter malicious content based on language patterns, suspicious attachments, and sender reputation.

- Endpoint security from multiple vendors with different malware detection engines further scrutinizes any emails that slip through.

- User awareness training empowers employees to identify and report suspicious emails, adding another layer of human intelligence.

This layered approach, where diverse solutions collaborate in real time, significantly impedes the attacker's progress. Instead of a smooth sail through a single point of entry, they encounter a treacherous ocean of constantly evolving defenses.

However, diversification isn't without its challenges. Managing a sprawling security ecosystem demands efficient integration, skilled expertise, and careful resource allocation. Integrating diverse solutions without introducing vulnerabilities or performance bottlenecks requires meticulous planning and ongoing optimization. Additionally, navigating the ever-expanding security landscape to choose the right tools for your specific needs requires constant vigilance and a willingness to adapt as threats evolve.

Despite these challenges, the dividends of a diversified defense far outweigh the risks. Consider the consequences of a major breach due to reliance on a single, outdated solution. By strategically layering diverse controls, you not only make it harder for attackers to succeed but also buy yourself invaluable time to detect and respond to threats.

Remember – DiD is an ongoing endeavor. Regularly assess your threat landscape, adapt your diversified defenses, and invest in tools and training that empower both your technology and your people. By embracing the chaos of complexity, you can transform your once-vulnerable castle into an impregnable fortress, leaving attackers lost in a labyrinth of your own making.

Layered security architecture

In the intricate tapestry of modern cybersecurity, layered security architecture emerges as the moon over a dark cloud, offering a structured approach to fortifying systems against a multitude of threats. This architectural paradigm involves the strategic implementation of multiple security layers, each acting as a barrier to potential threats and vulnerabilities. At its core, this approach aims to create a robust and cohesive defense framework by intertwining diverse security measures across different levels of an organization's infrastructure.

The traditional layered security approach offers a baseline defense, but against today's sophisticated threats, it's simply not enough. We need to move beyond stacking static layers and embrace resilient architectural patterns that seamlessly integrate diverse controls, maximizing their effectiveness and minimizing complexity.

Here are three powerful patterns for redefining the way we build secure systems.

Microservices with dynamic segmentation

Imagine a network segmented into tightly coupled microservices, each with its own robust defense perimeter. This is the essence of microservices with dynamic segmentation. By breaking down applications into fine-grained components, we limit the blast radius of an attack. Even if an attacker breaches one service, they're confined to its specific environment, unable to pivot laterally and access core assets.

Some common ways include the following:

- **Containerization**: Docker or Kubernetes containers provide isolation and granular control over individual services.
- **API gateways**: Secure access points enforce authorization and authentication for service communication.
- **Dynamic security policies**: Tools such as **Open Policy Agent** (**OPA**) [3] configure granular access controls based on context and service dependencies.

Benefits include the following:

- **Reduced attack surface**: Microservices minimize the target area for attackers, limiting potential damage.
- **Lateral movement restriction**: Dynamic segmentation confines breaches within affected services, preventing cross-service exploitation.
- **Improved recovery**: Isolating compromised services simplifies remediation and reduces downtime.

Zero trust architecture

Another pivotal architectural pattern gaining prominence is the zero trust architecture. This model operates on the premise of "never trust, always verify," advocating for continuous verification of user identities and devices attempting to access resources within the network. It challenges the traditional concept of implicit trust within the internal network and implements strict access controls and micro-segmentation, reducing the attack surface and minimizing lateral movement in the case of a breach:

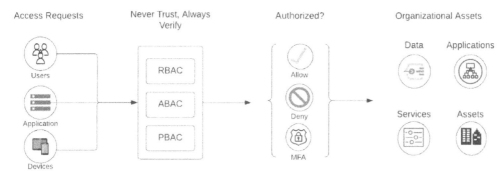

Figure 2.3 – Zero trust architecture

Some examples include the following:

- **Identity and access management (IAM)**: A centralized identity store such as **single sign-on (SSO)** [4] controls user roles and permissions across the fabric.

- **Zero trust security access service edge (SASE)**: Cloud-based SASE solutions dynamically evaluate context and identity before granting temporary access to authorized resources.

- **Data loss prevention (DLP)**: Tools such as Cisco Stealthwatch [5] or CrowdStrike Falcon [6] monitor data exfiltration attempts and enforce granular data access controls.

Benefits include the following:

- **Least privilege access**: Users only access the specific resources they need, minimizing the attack surface and potential impact.

- **Dynamic authorization**: Continuous context-aware access control reduces the risk of unauthorized lateral movement.

- **Improved visibility and control**: Centralized logs and real-time monitoring provide deep insights into user activity and potential threats.

Deception and advanced honeypots

Imagine hidden, instrumented systems scattered throughout your network, acting as bait for attackers. Advanced honeypots employ this strategy, deploying decoy environments and simulated data to mislead attackers and glean valuable intelligence. While not a primary defense, these tactics can buy time for response and reveal attacker techniques.

Some commonly deployed systems include the following:

- **Decoy systems**: Mimic production systems but contain no sensitive data, luring attackers to waste resources and expose their tools.

- **Behavioral honeypots**: Emulate specific applications or services to capture attacker techniques and tactics for analysis.

- **Threat intelligence (TI) integration**: Integrate honeypot data with TI feeds to proactively adapt defenses and identify emerging threats.

Benefits include the following:

- **Wasting attacker resources**: Decoy systems draw attackers away from real assets, providing valuable time for response.

- **Gathering TI**: Observing attacker behavior in controlled environments reveals their techniques and intentions.

- **Improved security posture**: Honeypot data informs risk assessments and drives proactive defense improvements.

These are just a few examples of how architectural patterns can redefine layered security. By embracing microservices, implementing zero trust principles, and deploying strategic deception, you can create a resilient digital landscape that leaves attackers confused and thwarted. Remember – security is a continuous journey. Constantly evaluate your architecture, adapt to new threats, and embrace innovative approaches. By proactively building resilience into your systems, you can stand strong against the ever-evolving threat landscape.

At a high level, these architectural patterns represent a shift toward proactive security, transforming how organizations perceive and implement their defense strategies. They address the limitations of singular security measures by weaving a complex network of interdependent controls and verification protocols across various layers of the system. Through these patterns, organizations achieve a heightened level of resilience, reducing the impact of potential breaches and empowering a more dynamic response to evolving threats.

Understanding and integrating these architectural patterns are crucial steps in designing robust and reliable security frameworks. They serve as blueprints, guiding organizations to construct systems fortified with diversified and interlocking defenses. Incorporating these principles into the fabric of cybersecurity architecture elevates the security posture, offering a comprehensive shield against the ever-evolving threat landscape.

DiD – Principles and benefits

Imagine hackers peering at your network, only to be met with a labyrinth of impenetrable layers. In this section, we'll dissect the core principles and undeniable benefits of the mighty DiD strategy. You'll learn to appreciate the fundamentals behind constructing a multilayered defense, thwarting attackers before they breach your most valuable assets. Let's dive into some of the core principles and benefits that can transform your security posture and ensure your data sleeps soundly every night.

Principles

- **Layering of defenses**: Implementing multiple layers of security controls at different levels to mitigate risks and reduce the likelihood of a successful breach.

- **Resilience through diversity**: Utilizing diverse security measures to counteract different attack vectors, minimizing vulnerabilities.

- **Redundancy and backup systems**: Deploying backup measures to ensure continuous operations and data integrity in case of primary system compromise. Don't rely on **single points of failure (SPOFs)**. Duplicate key controls to ensure breaches in one layer don't cripple overall security.

- **Principle of diminishing returns**: Recognizing that each additional security layer adds value but with diminishing marginal returns, emphasizing strategic investments.

- **Depth across all domains**: Extending defense across network, application, endpoint, and data layers to create a comprehensive security posture.

- **Proactive monitoring and response**: Employing continuous monitoring and rapid response mechanisms to detect and mitigate threats promptly.

- **Least privilege and segmentation**: Restricting access and segmenting networks to limit the impact of potential breaches and unauthorized access. Divide your network into isolated zones, restricting attacker movement and limiting the potential blast radius of breaches.

- **People-centric**: Empowering users with knowledge and best practices to recognize and thwart social engineering attempts. Remember – people are often the weakest link. Invest in security awareness training and empower employees to identify and report threats.

- **Adaptability and scalability**: Designing a flexible framework that adapts to evolving threats and business needs, scalable for future enhancements.

- **Continuous improvement and assessment**: Regularly evaluating and refining security measures to align with emerging threats and technological advancements.

- **Defense in layers, not stages**: Don't view these principles as a linear checklist. Consider them interwoven threads that combine and adapt to create a holistic, resilient security fabric.

Benefits

- **Increased resilience**: No single attack can easily penetrate your defenses, giving you time to respond and mitigate damage.

- **Reduced risk exposure**: Minimized impact of security incidents and lowered exposure to potential risks. Multiple layers make it harder for attackers to find vulnerabilities and exploit them.

- **Enhanced incident response (IR)**: Early detection and containment within specific layers limit the scope and impact of breaches.

- **Improved business continuity (BC)**: Redundancy and failover mechanisms minimize downtime and disruption to operations.

- **Heightened regulatory compliance**: Implementing a mature DiD strategy satisfies many security regulations and industry best practices.

- **Minimized impact of breaches**: Limiting the impact of breaches through isolated and contained security layers.

- **Reduced insurance costs**: Demonstrating a robust security posture can lead to lower premiums from cyber insurance providers.

- **Boosted customer and investor confidence**: Knowing your assets are well protected builds trust and fosters stronger relationships.

- **Competitive advantage**: Strong security attracts talent, protects innovation, and differentiates you from less secure competitors.

- **Sustainable security posture**: Building resilience into your core architecture lays the foundation for future adaptation and growth.

- **Cost-effective security**: Optimizing security investments by strategically distributing resources.

- **Adaptive security**: Adapting swiftly to new threats and technological advancements.

- **Holistic protection**: Comprehensive safeguarding of critical assets and sensitive information.

- **Peace of mind**: Knowing you've built a multilayered defense fortress fosters a sense of security and confidence within your organization.

By embracing these principles and leveraging the benefits they offer, you can create a truly defense-in-depth security posture that stands strong against ever-evolving threats, safeguarding your most valuable assets and securing a bright future for your business.

Remember – DiD is a journey, not a destination. Continuously improve your security posture, adapt to new threats, and trust the power of layered protection to keep your digital fortress unbreachable.

Security domains and controls

In the intricate landscape of cybersecurity, the concept of DiD unfolds across various security domains, each representing a crucial layer in the armor safeguarding organizations from an array of threats. DiD isn't merely throwing up a bunch of walls around your digital assets. It's constructing a labyrinthine fortress, where each domain acts as a distinct stronghold, interwoven with specialized controls to deter even the most persistent attackers. This section delves into the diversified realms of security domains, elucidating how their convergence orchestrates an impregnable defense strategy. Exploring these domains in tandem with their corresponding controls illuminates the pivotal role they play in fortifying the overarching security posture. It is your map to navigate this complex landscape, equipping you with the knowledge to build a layered and resilient security posture.

The journey commences by unraveling the intricate web of security domains – network, application, endpoint, data, and physical security. Each domain acts as a bastion, contributing its unique set of challenges and vulnerabilities, necessitating tailored controls to fortify its defenses. From network perimeters to internal data sanctuaries, these domains form the bedrock upon which robust defense architectures are constructed.

Diving deeper, the section unfurls an exhaustive array of controls, illuminating how each domain demands a distinct arsenal of defenses. Network security flourishes with firewalls, IDSs, and **virtual private networks** (**VPNs**), while endpoint security harnesses antivirus software, encryption, and user access controls. This comprehensive exploration equips you with a nuanced understanding of the specific measures essential to fortify different facets of their organization's security infrastructure.

This holistic comprehension fosters the ability to craft resilient defense architectures that repel diverse threats across the organizational spectrum. Embracing this knowledge cultivates the capacity to fortify systems comprehensively, mitigating risks and enhancing the overall security posture against the ever-evolving threat landscape. Let's dive in.

Mapping the landscape – Core security domains

The field of cybersecurity encompasses a vast and evolving landscape. To navigate its complexities and build a robust defense, practitioners often adopt the principle of domain-based DiD. This approach segments the security space into distinct functional areas, each focusing on a specific aspect of digital protection. In this section, we'll explore this crucial aspect of DiD by dissecting these key domains and examining how their specialized controls contribute to a layered and resilient security posture.

In the intricate world of cybersecurity, few things are as familiar as your online banking platform. Yet, beneath the surface of simple logins and transfers lies a complex tapestry of security domains, each meticulously woven to safeguard your financial trust. Let's embark on a journey through these domains, dissecting the journey of a simple request as it navigates from your fingertips to the bank's vaults.

Upon establishing a connection with the bank server through the **Transport Layer Security** (**TLS**) protocol over a **Transmission Control Protocol** (**TCP**) connection, a mutual determination between your system and the bank's server transpires to proceed or terminate the connection. This interaction occurs within the network as packets traverse the digital landscape. The network security domain encapsulates the digital boundaries, which include routers, switches, and gateways, delineating the organizational perimeter. This first line of defense filters out unwanted traffic, guarding the gateway to your financial fortress.

Conversely, the banking application you engage with integrates numerous controls aimed at safeguarding your information. The bank's web server extends beyond a singular entity running a unified application, often encompassing multiple microservices, external dependencies, and libraries. This collective security approach falls within the realm of application security, also recognized as product security. Strengthening application security entails rigorous code assessments, penetration testing, and adherence to secure coding practices. These measures fortify the system against potential loopholes and vulnerabilities that adversaries may exploit.

Servers function as machinery operating within a designated infrastructure, whether they're cloud-native applications or on-premises systems. Regardless of their environment, computational operations rely on hardware. The fortification of these machines holds equal importance to resolving vulnerabilities such as **cross-site scripting** (**XSS**) in a web application. This sphere falls under endpoint security, encompassing the establishment of controls aimed at thwarting, identifying, and addressing unwanted incidents that may occur on computational resources.

As previously mentioned, infrastructure stands as a pivotal component in ensuring the robust functionality of an application. Often underestimated, infrastructure serves as the powerhouse for all background operations, encompassing internal tools for development, testing, and deployment (commonly referred to as **continuous integration/continuous deployment** or **CI/CD**), as well as policies and controls to enforce secure software development practices. This critical realm is commonly known as infrastructure security.

The following diagram delineates various domains for implementing security controls, illustrating the application of a DiD strategy:

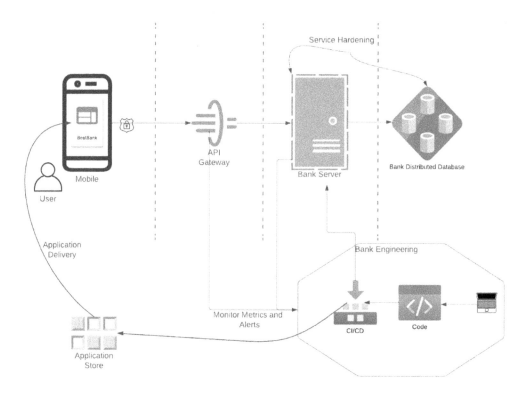

Figure 2.4 – Oversimplified defense strategy across domains

Information stands as the cornerstone of our digital requisites, anchoring the functionality of our systems. Within your banking application, data assumes a central role. Data security is dedicated to shielding sensitive information through encryption, stringent access controls, and resilient backup systems. These measures act as safeguards against data breaches, ensuring the preservation of confidentiality, integrity, and availability.

Consider a scenario where your bank's security team adeptly implemented multiple controls for each discussed domain but inadvertently overlooked securing the office premises, offering intruders direct access to hardware storing and processing sensitive data. In DiD, physical security is as crucial as its digital counterparts. Safeguarding premises, restricting access to critical infrastructure, and deploying surveillance systems bolster the protection of tangible assets. This holistic approach is integral to comprehensive security measures, ensuring a robust defense against potential breaches or unauthorized access.

In conclusion, your online banking experience is not just a simple click; it's a journey through a meticulously constructed labyrinth of security domains. Each domain, from the digital frontiers of network security to the physical solidity of physical security, plays a vital role in safeguarding your financial well-being. Understanding these domains and their importance is the first step toward appreciating the invisible armor that shields your banking trust, ensuring every transaction, every interaction, is as secure as your own vault.

Time to zoom in! Now that we've seen a real-world example and met the security domain crew, let's dissect each of their superpowers.

Network security

Network security concentrates on fortifying communication pathways and digital boundaries within an organization. It addresses vulnerabilities across networks, aiming to secure data transmission and access control and protect against external threats. By encrypting data flows, managing access points, and implementing firewalls, it aims to create a secure network environment. The domain's primary goal is to ensure the confidentiality, integrity, and availability of data traversing through the network infrastructure, thereby thwarting unauthorized access and potential cyberattacks.

Application security

Application security pertains to fortifying software and applications against potential threats and vulnerabilities. This domain aims to secure code, APIs, and software architectures from exploits and unauthorized access. By conducting stringent code reviews, penetration testing, and implementing secure coding practices, it endeavors to safeguard applications against potential attacks. Application security focuses on enhancing the reliability and integrity of applications, ultimately ensuring that software remains robust and resilient against emerging threats.

Infrastructure security

Infrastructure security is devoted to securing the fundamental hardware and software systems that power an organization's operations. It encompasses maintaining the reliability of internal tools, ensuring compliance with secure software development policies, and safeguarding the infrastructure supporting development, testing, and deployment processes. This domain concentrates on maintaining the availability, reliability, and resilience of the underlying technology stack, forming the backbone of a secure organizational infrastructure.

Endpoint security

Endpoint security centers on fortifying individual devices such as laptops, smartphones, and desktops, which serve as potential entry points for cyber threats. It aims to protect these endpoints from malware, unauthorized access, and data breaches. By deploying antivirus software, encryption mechanisms, and comprehensive access controls, it strives to secure endpoints against various forms of cyberattacks. Endpoint security is crucial in preserving the confidentiality, integrity, and availability of data stored or accessed via these devices.

Data security

Data security is primarily concerned with safeguarding sensitive information from unauthorized access, alteration, or destruction. It encompasses encryption, access controls, and robust backup systems to ensure the confidentiality, integrity, and availability of data. This domain focuses on mitigating data breaches and ensuring compliance with data protection regulations, thereby preserving the privacy and integrity of critical organizational information.

Physical security

Physical security emphasizes protecting physical assets, facilities, and infrastructure from unauthorized access, theft, or damage. It involves securing premises, employing surveillance systems, and implementing access controls to safeguard tangible assets. This domain aims to prevent unauthorized entry into facilities storing critical infrastructure, ensuring the physical safety and integrity of hardware and sensitive information. Physical security forms a vital layer in a comprehensive security framework, complementing digital security measures to create a holistic defense strategy.

Now that we have a better understanding of the values of each of the domains discussed, let's explore some commonly used controls.

Building the arsenal for each domain

Now, we stand amid the arsenal, where each security domain unveils its bespoke tools to combat digital threats. Network security boasts firewalls of varying strengths, IDSs as vigilant sentinels, and **deep packet inspection** (**DPI**) for meticulous scrutiny. Application security wields code reviews as microscopes, penetration testing for stress tests, and secure coding as fortifications. Endpoint security deploys antivirus shields, DLP gates, and encryption vaults to safeguard individual devices. Infrastructure security employs configuration management engineers, access control gatekeepers, and data backups as hidden caches. Finally, physical security relies on watchful eyes, access control checkpoints, and robust data center fortresses. This diverse arsenal, wielded by skilled practitioners, forms an intricate tapestry of defense that protects our digital kingdoms. We delve deeper into these controls in the following chapters, learning how to configure and implement them, ultimately forging an impregnable fortress against ever-evolving threats.

Network security controls

Network security has proven to be a battle-tested domain that has withstood the test of time. In the past few decades, we have witnessed significant growth and modernization in defensive controls across the network layer. In the following sections, we will delve into some commonly deployed controls within this domain.

Firewall

Talking about network security controls, the very first thing that pops up in mind is a firewall. Firewalls act as gatekeepers for networks, enforcing preconfigured security policies that govern incoming and outgoing traffic. They filter malicious connections and unauthorized activity, forming a crucial foundation for network security. Modern **next-generation firewalls (NGFWs)** take this a step further, offering deeper inspection capabilities to detect and block sophisticated malware and application-level attacks.

VPNs

A VPN acts like a secure tunnel on the internet, a secret passageway shielding your online activity from unwanted eyes. It encrypts your data, hiding sensitive information such as passwords and communications from prying eyes. Think of it as a digital bodyguard escorting you through the public internet, cloaking your true location by masking your IP address. This makes you practically invisible on risky public Wi-Fi networks, preventing eavesdropping and data theft. Bonus? You can even bypass geo-restrictions to access content unavailable in your area.

IDSs

IDSs operate as dedicated sensors for your network, continuously analyzing traffic flow for malicious activity. They employ a combination of signature-based and anomaly-based detection techniques. Signature-based approaches identify known attack patterns embedded in packet data, whereas anomaly-based methods flag deviations from established network traffic patterns and protocols. Upon detecting suspicious activity, IDSs generate alerts with detailed information such as source IP address, target system, and potential attack type. This timely notification empowers security personnel to investigate and respond swiftly, potentially mitigating threats before they compromise network integrity or sensitive data.

Network segmentation

Network segmentation partitions a physical network into logically isolated subnetworks, effectively creating internal boundaries. This granular approach enforces controlled traffic flow between segments, limiting the potential impact of security breaches or malware propagation. Implementing segmentation policies, such as **access control lists (ACLs)** and firewalls at segment boundaries, restricts unnecessary communication and elevates overall network security. By strategically dividing the network, segmentation strengthens the overall security posture, minimizing risk and enhancing the resilience of your digital infrastructure.

A very common network segmentation concept you might have come across in your cloud infrastructure is a **virtual private cloud (VPC)**. It acts as a secure enclave within the expansive shared infrastructure of a cloud provider, offering a customizable and isolated network environment for your cloud resources. [7]

Remember – the controls covered here represent a crucial slice of the network security arsenal, not the entire toolkit! Many other valuable tools and techniques exist, waiting to be explored and deployed to fortify your digital assets.

Application security controls

In the realm of application security, a vast array of controls exists, and we'll explore a select few within this section.

Threat modeling

As we covered threat modeling as an essential tool to build a secure system, let's revisit some core concepts quickly. Before lines of code are ever written, threat modeling steps in as a strategic architect for your application's security. Think of it as a blueprint for potential threats, meticulously identifying vulnerabilities and outlining defensive measures before attackers can exploit them. By analyzing data flows, user interactions, and system components, it pinpoints critical assets and maps out attack paths, empowering developers to proactively build security into the very fabric of the application. This proactive approach minimizes vulnerabilities, reduces remediation costs, and ultimately strengthens your application's resilience against real-world threats. If you're interested in building a framework for threat modeling, refer to the **Open Worldwide Application Security Project's (OWASP's)** detailed discussions [8].

Security testing

Proactive vulnerability discovery is the cornerstone of robust security. In this realm, security testing shines, offering a three-pronged approach: **static application security testing (SAST)**, **dynamic application security testing (DAST)**, and penetration testing.

SAST delves into the code's blueprint, meticulously analyzing its lines for flaws such as insecure APIs, SQL injection vulnerabilities, and XSS weaknesses. Imagine it as a code surgeon, dissecting the application's logic to identify potential security ailments before launch. Early detection through SAST helps developers build secure applications from the ground up.

DAST shifts focus to the running application, mimicking real-world attacks through fuzzing, SQL injection attempts, and automated exploit scanners. Think of it as a digital siege engine, relentlessly probing for vulnerabilities that might remain hidden in static analysis. DAST findings expose weaknesses in runtime behavior and deployed configurations, enabling security teams to patch before attackers leverage them.

Penetration testing, however, takes the gloves off. Skilled security professionals manually probe the application, deploying advanced hacking techniques to uncover deeper vulnerabilities that automated tools might miss. This "ethical hacking" exercise simulates real-world attacker behavior, revealing weaknesses in security controls, access controls, and overall system design. Penetration testing provides invaluable insights for hardening your defenses and closing critical security gaps.

Authentication and authorization

Authentication and authorization (often regarded as "AuthN and AuthZ") are two irreplaceable pillars of application security. Though we have covered these concepts thoroughly in *Chapter 1*, here's a quick refresher.

Authentication acts as the first line of defense, confirming who users are. It employs various methods such as passwords, MFA, and identity federation to validate user claims against trusted sources. Only once a user successfully proves their identity does the gate open. Authorization then takes the reins, deciding what a user can do within the application. It examines the user's role, permissions, and context to determine their access rights to specific resources or functionalities. For instance, a customer might be authorized to view order details but not edit pricing information.

Cryptography

It is debatable whether cryptography marked the rise of security. Personally, my journey in security began with this intersection of mathematics and security. In today's digital sphere, information travels far and wide, often traversing open networks vulnerable to prying eyes. This is where cryptography, the ancient art of secret writing, takes center stage. It acts as a powerful shield, transforming plaintext data into encrypted gibberish, unreadable without the proper key. Imagine a vault sealed with an intricate lock, accessible only to those with the authorized key.

Among its many tools, encryption stands as the cornerstone. Algorithms such as **Advanced Encryption Standard (AES)** and RSA scramble data using complex mathematical operations, rendering it useless to unauthorized actors. This encryption forms the backbone of secure communication protocols such as TLS. When you access a website using HTTPS, TLS kicks into gear. It establishes a secure tunnel between your browser and the web server, encrypting all data exchanged within that tunnel. This renders eavesdropping attacks futile, safeguarding sensitive information such as login credentials and financial data.

Cryptography and its tools such as encryption offer a vital layer of protection in today's interconnected world. By ensuring data confidentiality and integrity, they provide a secure foundation for trust to flourish.

Vulnerability rewards programs (VRPs)

Bug bounty programs act as crowdsourced vulnerability scouts, unlocking a global pool of security talent to relentlessly probe your applications for hidden flaws. By offering financial rewards for reported vulnerabilities, they incentivize ethical hackers to uncover weaknesses you might miss, empowering your security team to patch them rapidly and proactively bolster your defenses. Think of it as broadening your security team by including brilliant minds across the world who want to make the internet safer.

While not an exhaustive compilation of application security controls, it aims to spotlight commonly encountered ones in your day-to-day experiences.

Infrastructure security controls

In the world of enterprise security, anything that does not fall under network or application security often gets classified under infrastructure security. Let's explore some common controls we have at our disposal here.

Production systems hardening

Security often comes in the crosshair of usability. Making a system more secure often results in the system being less usable. Think of your ability to ship code to production without any barriers; while as a developer it might be a thrilling experience and beneficial to your development velocity, it is a security nightmare. Beyond security, it is a reliability concern as well, which we will not go into much detail on yet.

Reducing access to production systems often comes as one of the biggest infrastructure controls. The idea is to have no human altering states of production systems unilaterally. By reducing persistent access, you not only reduce the chances of accidental changes disrupting your organization but also limit what an attacker might be able to do even if they successfully impersonated an admin user.

CI/CD and supply chain security

At the heart of infrastructure security, securing the engine that builds and deploys your software is paramount. This is where CI security steps in, acting as a vigilant gatekeeper within the CI/CD pipeline. It safeguards every stage of the DevOps workflow, from code injection to configuration drift, ensuring only secure changes reach production. Through robust access controls, vulnerability scanning, and secrets management, CI/CD security builds a fortified pipeline, minimizing the risk of vulnerabilities and unauthorized access and guaranteeing the integrity and security of your deployed applications.

In addition to building and deploying software securely, it is paramount to understand and be aware of the organization's third-party dependency graph. Continuously monitoring external libraries and software becomes crucial as they are equally exploitable by attackers as your application code.

Continuous upgrades and patch management

While safeguarding the deployment pipeline is crucial, robust infrastructure security demands equal attention to the underlying computational resources that fuel it. Regardless of your organization's chosen platform – cloud or on-premises data centers – a proactive approach to system upgrades and security patching lays the cornerstone for a resilient security posture.

Endpoint security controls

From desktops to mobile phones, endpoint security controls stand vigilant, applying access controls, patching vulnerabilities, and detecting malicious activity across your fleet of devices, proactively defending your digital workforce and critical data.

Endpoint detection and response

Endpoint detection and response (EDR) solutions have become crucial for proactively identifying and combating threats on individual devices. EDRs operate deep within endpoints, continuously monitoring system activity, collecting telemetry data, and analyzing it for suspicious behavior. Armed with a combination of signature-based and anomaly-based detection techniques, they identify malicious processes, file changes, network connections, and registry modifications indicative of potential attacks. Upon detecting suspicious activity, EDRs generate detailed alerts, enabling security teams to rapidly investigate and respond. By isolating infected devices, quarantining malware, and rolling back changes, EDRs prevent breaches from spreading and minimize damage. However, EDRs remain limited to endpoint visibility, requiring additional solutions to provide a holistic view of the attack landscape. This is where **extended detection and response (XDR)** steps in.

XDR platforms integrate data from endpoints, networks, cloud environments, and **security information and event management (SIEM)** systems, providing a unified view of threat activity across an organization's entire infrastructure. This broader perspective empowers security teams to identify sophisticated attacks that might evade endpoint-only detection, correlate events across various sources, and prioritize response efforts efficiently. As cyber threats become increasingly complex and pervasive, EDRs and XDRs work together, offering critical layers of defense in the modern threat landscape.

Data leak prevention systems

Operating at the network and endpoint level, **data leak prevention systems (DLPs)** employ a combination of deep content inspection and user activity monitoring to detect the exfiltration or unauthorized sharing of confidential data. They leverage advanced techniques such as fingerprinting, pattern matching, and anomaly detection to identify sensitive data types, including **personally identifiable information (PII)**, **intellectual property (IP)**, and financial records. Upon discovering potential data leaks, DLPs initiate preconfigured actions, ranging from alerting security teams to blocking data transfers or encrypting sensitive content in transit. This multilayered approach minimizes the risk of accidental or malicious data exfiltration, safeguarding your organization's crown jewels from unauthorized access and ensuring compliance with regulatory data privacy mandates.

Unified endpoint management

With the rise of remote devices connecting to corporate networks, **unified endpoint management (UEM)** solutions are becoming prevalent. UEM platforms act as central hubs, consolidating and automating tasks such as device provisioning, configuration management, software and security patch distribution, and **application lifecycle management (ALM)**. By leveraging automation and real-time monitoring, UEM facilitates secure and consistent deployment of configurations and policies across diverse devices, simplifying IT operations and minimizing administrative overhead. Moreover, UEM integrates with security tools such as antivirus and EDR solutions, enabling centralized security policy enforcement and comprehensive endpoint protection. As organizations embrace hybrid work models and the **bring-your-own-device (BYOD)** trend flourishes, UEM emerges as the indispensable conductor, harmonizing device management, security, and user experience, orchestrating a secure and productive digital workplace.

Layering controls across security domains

The battle against cyber threats demands a multilayered defense, one that weaves together diverse controls across various security domains. Just as a medieval castle employed layered fortifications – moats, walls, gates, and guards – modern organizations must orchestrate defenses across network, application, endpoint, and data domains to effectively thwart attackers. Having examined various security domains and explored specific controls within each, let us reexamine the process of strategically layering controls across diverse domains to construct a robust and resilient security framework:

- **Securing the network perimeter**: At the outermost layer, network security controls act as vigilant gatekeepers, filtering incoming traffic before it reaches internal systems. Firewalls serve as the first line of defense, denying access to unauthorized destinations and protocols. **IDSs** and **intrusion prevention systems** (**IPS**) actively monitor network traffic for malicious activity, identifying suspicious patterns and blocking potential attacks before they reach applications or data. Additionally, network segmentation isolates critical assets within the network, limiting lateral movement and minimizing potential damage in case of a breach.

- **Hardening the application**: Moving inward, application security controls shield the software that processes and stores sensitive data. SAST analyzes code for vulnerabilities before deployment, while DAST simulates real-world attacks to uncover runtime weaknesses. **Web application firewalls** (**WAFs**) act as gatekeepers for specific applications, filtering malicious requests and protecting against common web attacks such as SQL injection and XSS. Access controls and authorization mechanisms further restrict access to sensitive functionalities and data within applications.

- **Protecting the endpoints**: Beyond the network and application layers lie the individual devices used by employees. Endpoint security controls such as antivirus and anti-malware software provide real-time protection against known threats. EDR solutions continuously monitor endpoints for suspicious activity and enable rapid response to potential breaches. DLP systems prevent unauthorized data exfiltration, safeguarding sensitive information across desktops, laptops, and mobile devices. Encryption further bolsters endpoint security by protecting data at rest and in transit, ensuring its confidentiality even if a device is compromised.

- **Protecting the data**: The innermost layer of the digital castle holds the crown jewels – sensitive data itself. Data security controls such as encryption of databases and filesystems safeguard data at rest, rendering it unreadable without the proper key. Activity monitoring and logging track user access and data modifications, aiding in anomaly detection and forensic investigations. Data governance policies and procedures dictate how data is accessed, used, and protected, ensuring compliance with relevant regulations and minimizing the risk of unauthorized access or misuse.

By strategically layering controls across diverse domains, organizations construct a resilient and multifaceted defense posture. Each layer acts as a barrier, mitigating threats and slowing down attackers, buying precious time for security teams to detect and respond. This layered approach ensures that even if attackers breach one layer of defense, they face formidable challenges in penetrating subsequent layers, thereby significantly enhancing an organization's overall security posture.

Selecting and implementing the right controls

After a careful review of different security domains and the controls each of them has to offer, in this section, we delve into the pivotal aspect of assembling a comprehensive DiD strategy by carefully selecting and integrating security controls across various domains. We navigate the intricate process of choosing the most effective controls within each domain and strategically layering them to fortify an organization's security posture.

Securing your digital realm against ever-evolving cyber threats demands a strategic selection of security controls. Implementing a diverse array of tools across network, application, endpoint, and data domains is crucial, but haphazard control selection can lead to redundancies, gaps, and inefficiencies. This section equips you with a practical framework for choosing the right controls to build a robust DiD strategy.

So far in this chapter, we have moved beyond simply listing best-practice controls and delved into a systematic approach for evaluating your specific needs and assessing potential threats. Now, let's shift our focus toward prioritizing the most effective controls for each security domain. Through this framework, you'll gain the ability to do the following:

- Identify critical assets and vulnerabilities within each domain.

- Map potential threats and attack vectors relevant to your organization.

- Evaluate the capabilities and benefits of various security controls.

- Prioritize and optimize control selection based on cost, efficacy, and integration.

- Build a layered and cost-effective DiD strategy tailored to your unique security posture.

By following this framework, you'll transform from a passive recipient of security recommendations into an active architect of a comprehensive and adaptable defense, one meticulously constructed to withstand the ever-shifting winds of the cyber landscape.

Assessment of organizational needs

Before embarking on the journey of meticulously selecting security controls, understanding your organization's unique needs and priorities is an absolute necessity. This foundational assessment acts as a roadmap, guiding you toward cost-effective and impactful control implementation.

One crucial aspect of this assessment involves diligently **identifying and classifying your critical assets**. This encompasses not just sensitive data such as financial records and customer information but also key infrastructure components, applications, and intellectual property. Employing techniques such as data inventories, system mapping, and stakeholder interviews, pinpoint the elements vital to your business operations and data integrity. Categorize these assets based on their criticality, potential impact of compromise, and regulatory compliance requirements. This prioritization exercise forms the bedrock of your security investment decisions. One common mistake I have seen is to focus extensively on the data and forget other critical internal components such as encryption keys, and so on.

Next, delve into the world of **threat assessment**. Analyze threats relevant to your specific industry, size, and geographic location. This involves understanding common attack vectors, emerging cyber trends, and vulnerabilities associated with your technology stack and business processes. Conduct vulnerability scans and penetration testing exercises to uncover potential weaknesses and assess the likelihood of exploitation. By mapping threats against identified critical assets, you gain invaluable insights into which assets face the greatest risk and require the most robust protection. The key idea is to make the process of threat modeling easy and repeatable. By doing that, you ensure it is reusable and adaptable. The threat landscape for an organization is always evolving, and the best defense strategy is continuous.

Finally, remember that security must be viewed through the lens of business value. Each control carries its own implementation and maintenance costs, which must be weighed against the potential losses resulting from a security breach. Conduct **cost-benefit analyses (CBAs)** for various control options, factoring in factors such as data sensitivity, regulatory fines, and reputational damage. This ensures that your security investments yield optimal value and align with your overall business objectives.

By diligently and meticulously engaging in these foundational exercises, you lay the groundwork for selecting the most effective and cost-efficient security controls. With a clear understanding of your critical assets, relevant threats, and business priorities, you can navigate the complex landscape of security solutions with confidence, crafting a DiD strategy tailored to your organization's unique needs. Next, let's delve deeper into how unique organizational threats dictate its defense programs.

Matching controls to threats

Matching the right security controls to specific threats lies at the heart of building a resilient DiD strategy. This process extends beyond simply identifying vulnerabilities and plugging the gaps with readily available tools. It requires a nuanced understanding of the broader context in which your organization operates, considering factors such as regulatory compliance, customer trust, BC, and the nature of potential adversaries.

Regulatory compliance

Firstly, regulatory mandates and industry best practices play a crucial role in shaping your control selection. Compliance requirements from entities such as the **Health Insurance Portability and Accountability Act (HIPAA)** or the **Payment Card Industry Data Security Standard** (PCI DSS) often dictate specific security controls across data protection, access management, and logging practices. Beyond security controls, companies need to deliberately pay attention to privacy compliances such as the **General Data Protection Regulation** (GDPR), the **California Consumer Privacy Act (CCPA)**, and so on. Failing to implement these mandatory controls can result in hefty fines and reputational damage. Identifying relevant regulations and aligning your control selection accordingly ensures legal and ethical compliance while minimizing risk.

User trust

Beyond compliance, protecting customer trust demands proactive defense against attacks that could compromise sensitive data or disrupt essential services. Data breaches or system outages can erode customer confidence and loyalty, impacting your brand reputation and bottom line. By prioritizing controls that safeguard customer data and ensure service availability, you build trust and demonstrate a commitment to their security. *Target* reported a 46 percent profit plunge in the fourth quarter of 2013 due to a massive financial data breach [9].

Furthermore, consider the potential business disruptions caused by successful attacks. A ransomware attack disrupting critical operations or a supply chain compromise halting production can lead to significant financial losses and operational downtime. Assessing the potential impact of different threats enables you to prioritize controls that mitigate the most disruptive scenarios and ensure BC.

Threat actors

Finally, the nature of potential adversaries should also influence your control selection. A small-scale cybercriminal might be interested in stealing financial data, while a nation-state actor might target your IP or critical infrastructure. For instance, if your company's software runs on more than a billion devices, it can be a very attractive target for **advanced persistent threats** (**APTs**) or nation-states. Understanding the capabilities and motivations of potential attackers allows you to implement targeted controls that effectively deter and thwart their objectives.

By carefully considering these multifaceted factors, you move beyond a reactive approach to security and embrace a proactive strategy that aligns controls with specific threats. This ensures your DiD strategy is tailored to your unique context, effectively safeguarding your critical assets, customers, and business operations against a broad range of potential adversaries.

Control selection criteria

With a clear understanding of your organizational needs and the threats your assets face, now let's navigate the complex landscape of control selection. Choosing the right tools requires meticulously evaluating them against a set of key criteria, ensuring they effectively address your security needs while remaining practical and sustainable for your unique environment.

Efficiency

One of the core criteria for selecting a security control is its effectiveness. Each control must demonstrably mitigate the specific threats targeted. Evaluating control efficacy involves analyzing their capabilities in terms of detection, prevention, or containment of specific attack vectors and vulnerabilities. Conduct research, consult experts, and leverage TI reports to ensure your chosen controls possess the appropriate countermeasures against relevant threats.

Cost

However, effectiveness alone is not enough. Cost considerations play a crucial role in any security investment. Carefully analyze the acquisition, deployment, and maintenance costs associated with each control. This includes capital expenditures for software and hardware, ongoing subscription fees, training for personnel, and operational overhead. Choose controls that deliver optimal efficacy within your budgetary constraints, prioritizing high-impact threats and avoiding costly solutions for marginally beneficial gains. This goes back to an idea presented in *Chapter 1*; you do not protect a $100 asset with $1M security control.

Integration with the existing stack

Beyond cost, integration complexity should also be evaluated. Seamless integration with existing infrastructure is vital for efficient deployment and minimal disruption. Consider factors such as compatibility with existing systems, ease of configuration, and resource utilization before selecting controls. Opt for solutions that integrate smoothly with your current environment, minimizing implementation headaches and downtime.

Scalability and flexibility

As with any software system, a robust DiD strategy demands scalability and flexibility. As your organization evolves, security controls must adapt to accommodate changing needs and emerging threats. Prioritize controls that can be readily scaled to accommodate future growth, support new technologies, and adapt to new vulnerabilities. Avoid rigid solutions that may quickly become obsolete or require costly overhauls as your security landscape transforms.

Compliance

As we covered in an earlier section, regulatory compliance is another input in organizational control selection. Aligning your control selection with relevant regulations and industry standards ensures legal and ethical compliance while potentially reducing risk and demonstrating your commitment to data security. Identify applicable regulations such as HIPAA or GDPR and choose controls that demonstrably meet their requirements, simplifying compliance audits and safeguarding against potential non-compliance penalties.

By thoroughly evaluating controls against these critical criteria, you move beyond a chaotic approach to a strategic and informed selection process. You build a DiD strategy tailored to your specific needs, balancing effectiveness, cost, integration, scalability, and compliance to create a resilient and sustainable security posture for your organization.

Implementation strategies and best practices

Choosing the right security controls is only half the battle. Seamlessly integrating them into your existing infrastructure and ensuring their effectiveness requires meticulous planning and a focus on best practices. Now, we will dive into the crucial implementation phase, outlining strategies for maximizing the impact of your selected controls:

- **Prioritizing integration**: Successful implementation of a security control lies in flawless integration with your existing systems and security architecture. Prioritize controls that offer flexible deployment options, seamless integration with current tools and platforms, and minimal disruption to ongoing operations. Conduct feasibility assessments and compatibility testing before deployment, eliminating potential integration issues that could hinder effectiveness or disrupt critical workflows.

- **Piloting for efficiency**: Large-scale rollouts can be fraught with unforeseen challenges. To mitigate risks and ensure smooth adoption, pilot implementations in controlled environments are invaluable. By testing a control on a smaller scale, you gain valuable insights into its functionality, integration nuances, and user experience. This allows you to identify and address any issues before a widespread deployment, minimizing disruption and optimizing efficiency. The importance of piloting a large-scale change grows proportionally with the size of your organization. The deployment of security controls should occur with minimal, if any, service interruptions to prevent financial losses and reputational damage to the organization.

- **Layering for complementarity**: Remember – DiD thrives on synergy. Interoperability and communication between different control layers are crucial for maximizing their collective effectiveness. Choose controls that offer robust integration capabilities, allowing them to share TI, trigger coordinated responses, and seamlessly escalate potential incidents across various security domains. To avoid having duplicate controls for the same threat, it is important to pay close attention to existing controls and how a new one could complement them.

- **Communication and training**: Successful implementation extends beyond technical considerations. Engaging in clear communication and comprehensive training for your team is crucial for user adoption and effective control utilization. Explain the rationale behind implementing specific controls, educate users on their functionalities and best practices, and provide ongoing support to ensure proper usage and adherence to security protocols. One example to highlight here is to train your software development teams to perform primary rounds of threat modeling on their services and applications. Transferring some of these skills to development teams is scalable and can result in faster cycles.

- **Continuous monitoring and optimization**: Remember – implementation is not a one-time event. Continuous monitoring and optimization are essential for maintaining the effectiveness of your security controls. Regularly evaluate control performance, analyze logs and reports, and adapt configurations as needed to address evolving threats and user behavior. We will take a deeper look into monitoring in the next section.

By focusing on these best practices, you transform the implementation phase from a technical hurdle into a strategic opportunity to optimize your DiD strategy. By prioritizing integration, utilizing pilot deployments, fostering interoperability, engaging your team, and embracing continuous improvement, you ensure your chosen controls reach their full potential, safeguarding your critical assets and building a resilient security posture for your organization.

Continuous monitoring and adaptation

The digital landscape is constantly evolving, with new threats emerging and business needs shifting. Therefore, the final and arguably most crucial piece of the puzzle lies in continuous monitoring and adaptation of defense strategies. This ongoing process ensures your security controls remain relevant, effective, and aligned with your evolving risk landscape.

Regular assessment and monitoring of your security controls is non-negotiable. Analyze logs and reports to identify suspicious activity, assess the effectiveness of deployed controls against targeted threats, and evaluate whether control configurations require adjustments. Conduct vulnerability scans and penetration testing exercises at regular intervals to uncover potential weaknesses and proactively address them before they can be exploited. This continuous evaluation ensures your defenses stay sharp and adapt to the ever-shifting threat landscape.

Metrics and **key performance indicators** (**KPIs**) play a vital role in this iterative process. Define metrics that track the performance of your controls, such as detection rates, IR times, and blocked attacks. By monitoring these metrics and analyzing trends, you gain valuable insights into the effectiveness of your security posture and identify areas for improvement. Additionally, track user behavior and system activities to identify anomalies and potential threats that might not be easily detectable through traditional security means. This data-driven approach allows you to make informed decisions about control adjustments, resource allocation, and future investments in the security arsenal.

Finally, IR serves as a crucial feedback loop in the continuous monitoring and adaptation process. Analyze past incidents to understand how attackers breached your defenses, which controls failed to prevent the attack, and what vulnerabilities were exploited. Use these insights to refine your security controls, update configurations, and implement additional safeguards to prevent similar attacks in the future. This closed-loop feedback ensures your DiD strategy learns from past experiences and evolves to become increasingly resilient against future threats.

By implementing continuous monitoring, designing flexible defense mechanisms, and through data-driven analysis, you transform your security posture from a static shield into a dynamic and responsive bulwark. This iterative approach ensures your controls remain relevant and effective in the face of evolving threats, safeguarding your critical assets and keeping your organization secure in the ever-changing digital landscape.

Glimpse of a real-world DiD approach

Having explored a myriad of concepts across various security domains and learned how to construct a layered defense strategy, let's now put that knowledge into practice by designing defensive controls against one of the most common vulnerabilities found in modern web applications.

Threat

Server-side request forgery (**SSRF**) occurs when an attacker takes advantage of a server-side flaw to coerce it to fetch resources and access or manipulate information locally in the server context that the attacker should not have access to [10].

SSRF allows attackers to manipulate a web application into sending unauthorized requests from the server. This exploit typically occurs when an application processes user-supplied URLs to fetch resources. Attackers can abuse this vulnerability to access internal systems, pivot to other parts of the network, or execute attacks against different servers, bypassing access controls and posing significant security risks to the affected system.

Impact

SSRF is very common due to the increasing functionalities of modern web applications. Exploiting an SSRF vulnerability can result in severe repercussions for an organization. Attackers can abuse this flaw to bypass firewalls and access internal systems, potentially retrieving sensitive information or executing commands on behalf of the affected server. This could lead to data breaches, unauthorized access to sensitive resources, and compromise of critical infrastructure. Additionally, SSRF can facilitate attacks on other systems within the network, amplifying the risk and impact of the exploit.

Mitigation

Let's take a DiD approach to mitigating this risk. The following strategy is heavily influenced by OWASP Top 10 [11].

Network security controls

- Isolate sensitive servers and applications within separate network segments to contain any potential SSRF exploits and limit their reach.

- Enforce strict firewall rules ("default deny") to block outgoing connections to unauthorized domains or IP addresses, narrowing the potential attack surface.

Application security controls

- Rigorously validate and sanitize all user-controlled input, including URLs, parameters, headers, and cookies, ensuring they conform to expected formats and don't contain malicious elements.

- Define explicit allow lists of authorized resources the server can access, preventing connections to untrusted domains. Enforce allow listing at both the network and application levels for robust protection.

- Sanitize responses before sending them to clients and disable HTTP redirection.

- Regular internal or external penetration testing can proactively identify and remediate weaknesses before they can be exploited.

Infrastructure security controls

- Regularly update web servers, application frameworks, and libraries to patch known SSRF vulnerabilities, addressing common attack vectors [12].

- Minimize the attack surface by disabling any services or features that aren't essential for server functionality. Move security-critical applications from external networks [13].

The aforementioned defense strategy serves as a valuable guide and offers specific control recommendations across different security domains. However, the optimal implementation will vary depending on individual organizational needs, infrastructure, and risk profiles. Organizations should adapt and tailor this framework to their specific contexts to achieve the most effective protection against SSRF attacks.

Summary

This chapter, *Practical Guide to Defense in Depth*, expanded on the risk-based security strategy introduced earlier by diving into the multilayered approach to security. This chapter emphasized the importance of security across various domains (network, application, data, and so on) and highlighted the arsenal of available security controls within each.

We covered in-depth knowledge of these controls, empowering you to strategically select and integrate them into your organization's security framework. In essence, this chapter equips you with the tools and understanding to translate a risk-based security strategy into a tangible, layered defense for your organization.

Key takeaways

- Security is a chain, and the weakest link breaks it.

- A defender needs to be successful every time; an attacker just needs to get lucky once.

- Each defense layer makes it more complex and expensive for an attacker to break in.

- Implementing multiple layers of security controls at different levels mitigates risks and reduces the likelihood of a successful breach.

- We've explored how a multilayered approach, encompassing diverse security domains such as network, application, and data, significantly strengthens your overall security posture.

- You've gained valuable insights into the vast array of security controls available within each domain, equipping you to make informed decisions when building your defensive arsenal.

- This chapter emphasized the significance of carefully selecting and integrating controls into your specific organizational context and risk profile. Remember – there's no one-size-fits-all approach; tailoring is key!

- By strategically layering and integrating diverse controls, you've learned how to construct a dynamic and resilient defense against ever-evolving threats, safeguarding your valuable assets and ensuring BC.

Equipped with a DiD mindset and the bedrock principles behind it, next, we will explore how to build a useful framework to transform organizational security into a layered strategy.

Congratulations on getting this far – pat yourself on the back! Good job!

Further reading

To learn more about the topics that were covered in this chapter, look at the following resources:

- [1] Attack surface definition: `https://en.wikipedia.org/wiki/Attack_surface`

- [2] Twitter attack: `https://blog.twitter.com/en_us/topics/company/2020/an-update-on-our-security-incident`

- [3] OPA: `https://www.openpolicyagent.org/`

- [4] SSO: `https://en.wikipedia.org/wiki/Single_sign-on`

- [5] Cisco Stealthwatch: `https://www.cisco.com/c/en_hk/products/security/stealthwatch/index.html`

- [6] CrowdStrike Falcon: `https://www.crowdstrike.com/products/data-protection/`

- [7] Network segmentations (VPCs): `https://maturitymodel.security.aws.dev/en/2.-foundational/vpcs/`

- [8] OWASP threat modeling cheatsheet: `https://cheatsheetseries.owasp.org/cheatsheets/Threat_Modeling_Cheat_Sheet.html`

- [9] Target breach of 2013: `https://www.nytimes.com/2014/02/27/business/target-reports-on-fourth-quarter-earnings.html`

- [10] SSRF definition: `https://en.wikipedia.org/wiki/Server-side_request_forgery`

- [11] OWASP Top 10: `https://owasp.org/Top10/A10_2021-Server-Side_Request_Forgery_%28SSRF%29/`

- [12] Instance Metadata Service Version 2 (IMDSv2) upgrade: `https://aws.amazon.com/blogs/security/defense-in-depth-open-firewalls-reverse-proxies-ssrf-vulnerabilities-ec2-instance-metadata-service/`

- [13] DiD – network security in the cloud: `https://cloud.google.com/blog/products/networking/google-cloud-networking-in-depth-three-defense-in-depth-principles-for-securing-your-environment`

3
Building a Framework for Layered Security

So far, we've navigated the intricacies of risk-based security and delved into the foundational building blocks of **Defense in Depth (DiD)**. Carrying our learnings along the way, we reach the pivotal moment: constructing a framework that translates these principles into tangible action items. This chapter serves as your blueprint, guiding you in creating a security strategy capable of withstanding the relentless tide of evolving threats.

Our journey begins with establishing a robust framework anchored in the knowledge from earlier chapters. We'll dissect the components of this framework, examining how risk assessments inform control selection, how security domains synergistically bolster defenses, and how continuous monitoring ensures perpetual vigilance. This blueprint, tailored to your organization's unique landscape, will serve as the foundation upon which your layered security strategy is built.

But a resilient edifice demands more than just sturdy foundations. Consistent application across organizational boundaries is paramount. Here, we'll explore the role of security policies and compliance in weaving a web of coherence throughout your structure. We'll delve into how harmonizing disparate elements fosters consistency, simplifies processes, and ultimately contributes to a more robust security posture.

Yet, even the most meticulously crafted framework remains inert without enforcement and accountability. This chapter tackles these crucial pillars, outlining techniques for effective implementation, monitoring adherence, and fostering a culture of security responsibility. You'll discover practical strategies for empowering your team, building buy-in, and ensuring every member plays their part in shielding your organization against evolving cyber threats.

As you embark on this chapter, remember that this is not merely a technical exercise; it's a journey toward solidifying your organization's digital defenses. By mastering the art of constructing a layered security framework, you empower yourself to face the future with confidence, knowing your data, systems, and operations are shielded by a bastion of resilience, capable of enduring the ever-shifting challenges of our digital world.

In this chapter, we're going to cover the following main topics:

- Establishing a robust framework
- Consistency and standardization by security policies
- Compliance and regulatory requirements
- Enforcement and accountability

Let's get started!

Establishing a robust framework

NIST defines DiD as an *"information security strategy integrating people, technology, and operations capabilities to establish variable barriers across multiple layers and missions of the organization"* [1]. A comprehensive DiD strategy is forged at the convergence of an organization's human resources, technological infrastructure, and operational practices. Throughout this chapter, we will dissect this convergence and outline essential principles that form the backbone of our framework.

A resilient security framework is crucial in safeguarding digital environments amid ever-evolving threats. This section delves into constructing an effective DiD strategy, amalgamating insights from risk-based security paradigms and the core tenets of DiD. The framework's architecture weaves together administrative, physical, and technical controls, reinforcing the layers of defense across different domains. Anchored by principles of consistency, compliance, and validation, this framework evolves in tandem with the dynamic threat landscape, fostering resilience and adaptability in modern security strategies.

At its core, this section seeks to demystify the intricate tapestry of security measures, stitching together a comprehensive blueprint that transcends isolated controls. By navigating through the intricacies of security policies, compliance mandates, and the execution of accountability-driven strategies, this framework aims to empower security practitioners with actionable methodologies to fortify organizational defenses. It aims to illuminate the path toward constructing an adaptable, comprehensive security infrastructure that resonates with the evolving demands of the digital ecosystem.

Organizing defensive controls

As we embark on building a robust layered security framework, it's crucial to recognize that effective defense isn't solely reliant on technological might. We must weave a diverse tapestry of controls, encompassing not only the realm of technology but also the vital threads of administrative and physical measures. This comprehensive approach, known as DiD, ensures that attackers face a layered gauntlet of obstacles at every turn, significantly increasing the complexity and cost of breaching your defenses.

Let's untangle the strands of this protective tapestry. Administrative controls form the foundational layer, shaping your organization's security culture and behavior. Think of them as the carefully crafted rules and procedures that guide how your people handle sensitive information and access critical systems. These include security policies, user training programs, incident response plans, and access control protocols. By establishing clear expectations and fostering security awareness, administrative controls lay the groundwork for a vigilant and responsible organization.

But technology plays a crucial role too. Technical controls serve as the technological fortifications within your layered defense. Firewalls, intrusion detection systems, encryption technologies, and endpoint security solutions form the technological bulwarks, actively monitoring for threats, blocking unauthorized access, and safeguarding sensitive data. By strategically selecting and deploying these controls, we erect a digital barrier between your assets and potential attackers.

Finally, physical controls serve as the tangible guardians of your physical infrastructure. Think of them as the locked doors, security cameras, and access control systems that restrict physical access to your critical systems and data centers. These measures complement the digital defenses by acting as a final line of defense against physical intrusion attempts.

By recognizing the value of each thread – administrative, technical, and physical – and weaving them together into a coherent whole, we create a layered security framework of unparalleled resilience. Each layer complements the others, creating redundancies and increasing the attacker's effort and expense should they attempt to breach your defenses. Remember, true security lies not just in individual controls, but in their synergistic interplay, forming an unbreakable shield against the ever-evolving threats of the digital world.

Administrative controls

While firewalls and intrusion detection systems often steal the spotlight in security discussions, let's not underestimate the potent force of administrative controls. These unsung heroes form the bedrock of any resilient security framework, wielding the power of human behavior and organizational practices to significantly reduce the attack surface and foster a culture of security awareness.

Think of administrative controls as the carefully crafted rules of engagement – the guiding principles that dictate how your organization interacts with technology and sensitive information (to know more, you can read "Constitutional AI" from Anthropic [2]). They go beyond mere technical configurations, shaping behavior and instilling best practices into the very fabric of your daily operations. Security policies, for instance, act as the constitution of your security posture, outlining acceptable use, access privileges, and incident response procedures. These policies, coupled with rigorous training programs, empower your employees to become active participants in maintaining a secure environment. By equipping them with the knowledge and tools to identify and report suspicious activity, you leverage the vast human sensor network within your organization, turning every employee into a vigilant defender.

You might be recalling those yearly security training sessions at your organizations; it's a form of administrative control. While defining policies alone cannot protect your digital assets, it becomes the mobilizer of the enforcement. Imagine your company has a policy to ensure two-party review for any changes in production. Now it becomes much more straightforward for a security team to implement defense mechanisms that restrict unilateral state-changing operations in production.

Administrative controls extend beyond simply establishing rules and conducting training. Incident response plans, for example, serve as the playbook for navigating security breaches, ensuring swift and coordinated action in the face of an attack. They define roles and responsibilities, communication protocols, and recovery procedures, minimizing potential damage and facilitating a swift return to normalcy.

> **Remember**
>
> Even the most sophisticated technical defenses can be thwarted by human error or negligence. By investing in robust administrative controls, you cultivate a security-conscious culture, empower your employees to become defenders, and build a resilient human firewall that complements and strengthens your technological fortifications.

In a nutshell, it's the seamless interplay of both human and technical elements that truly defines a formidable security posture. Humans often are considered as the weakest link in any security strategy. While this is largely due to the negative effects of targeted phishing campaigns resulting in data breaches, I'd argue it's a sign of weak administrative controls at an organization. Though humans remain the primary target for attacks, empowering them through administrative measures equips them to actively contribute to the overall defense strategy.

Technical controls

As we ascend the layered citadel of DiD, we reach the imposing walls of technical controls. These digital bastions form tangible defenses, actively monitoring, filtering, and blocking threats throughout your network and devices. Unlike their administrative counterparts, focused on shaping human behavior, technical controls wield the power of hardware, software, and network configurations to directly intervene and thwart malicious activity.

Software controls stand as vigilant sentries within your systems. Antivirus and anti-malware solutions scan for malicious code, **intrusion detection systems** (IDSs) monitor network traffic for suspicious patterns, and **data loss prevention** (DLP) tools safeguard sensitive information by restricting its unauthorized movement. These software warriors act as the first line of defense, identifying and neutralizing threats before they can inflict damage.

Hardware controls form the physical foundation of your digital fortress. Firewalls act as gatekeepers, meticulously scrutinizing inbound and outbound network traffic and allowing only authorized communication to pass through. **Hardware security modules** (HSMs) [3] perform encryption to scramble sensitive data, rendering it unreadable to anyone without the decryption key and significantly increasing the attacker's effort and potential cost of accessing it. Access control systems, whether physical or digital, restrict access to critical systems and data centers, ensuring only authorized individuals can gain entry. With the advancement of attack strategies, as an industry, we are moving toward hardware modules guaranteeing security at the boundary, as we are seeing a rise in the adoption of HSMs for critical encryptions, and **Trusted Platform Modules** (**TPMs**) [4] to establish root of trust.

The concept of the reference monitor [5]

In the realm of computer security, a reference monitor is an abstract concept that describes a hypothetical, tamper-proof mechanism responsible for enforcing system security policies.

Key properties are as follows:

- **Non-bypassable:** The reference monitor must intercept and mediate every access attempt on protected resources. No access can bypass its scrutiny, ensuring comprehensive protection.
- **Tamper-proof:** The reference monitor must be protected from unauthorized modification or interference, even by privileged system components. This guarantees its integrity and trustworthiness in enforcing security policies.
- **Verifiable**: The reference monitor's operations must be open to scrutiny and verification, ensuring its correct implementation and adherence to security policies. This promotes transparency and trust in the system's security mechanisms.

TPMs are hardware-based security chips that strive to embody the reference monitor concept in real-world computing systems. They provide a tamper-resistant environment for storing sensitive data, generating cryptographic keys, and enforcing security policies.

While TPMs don't perfectly replicate the ideal reference monitor, they offer a practical approximation, significantly enhancing the security of computing systems. By providing a trusted foundation for enforcing security policies, TPMs play a crucial role in safeguarding sensitive data and thwarting unauthorized access attempts.

On the other hand, **network controls** bind these diverse elements together, weaving a secure tapestry across your infrastructure. Network segmentation compartmentalizes your network, creating isolated zones that limit the spread of potential infections. **Security information and event management** (**SIEM**) systems gather and analyze data from multiple security tools, providing a holistic view of your security posture and identifying emerging threats. The rise of machine learning gave rise to powerful behavior analysis engines such as **user and entity behavior analytics** (**UEBA**) tools to detect anomalies within collected event logs.

Remember, technical controls are not standalone fortresses. Their true power lies in their synergistic interplay. By layering and integrating these diverse elements – software, hardware, and network controls – you create a sophisticated defense that confuses, slows down, and ultimately frustrates even the most advanced attackers. Each layer acts as a hurdle, increasing the time and effort required for a successful breach.

Physical controls

While technology forms the digital shield of DiD, let's not underestimate the vital role of physical controls. These are the vigilant guards patrolling your physical infrastructure, deterring and detecting unauthorized intrusions, and offering the first line of defense against intruders with screwdrivers instead of keyloggers.

Imagine your data center as a high-security vault. Physical controls function as imposing locks, security cameras, and alarm systems that secure its perimeter and deter unauthorized entry. From sturdy fences and controlled access points to meticulous visitor logging and CCTV monitoring, these measures make it clear that your sensitive assets are not lightly guarded. They act as a physical obstacle course, forcing potential intruders to expend time, resources, and risk exposure before even attempting to bypass your digital defenses.

Remember, even the most sophisticated cyber defenses can be thwarted by physical access to critical infrastructure. By investing in strong physical controls, you buy yourself precious time to respond to intrusions, activate countermeasures, and minimize potential damage. These guardians stand shoulder-to-shoulder with their technical counterparts, creating a layered and cohesive security posture that protects your valuable assets in both the digital and real worlds.

Administrative, technical, and physical controls constitute crucial elements for establishing a comprehensive layered security program. While each control type offers distinct advantages, their synergistic interaction enhances the system's resilience against even the most sophisticated attackers. The accompanying visual representation illustrates this integration, depicting the three control types as integral slices within a circle:

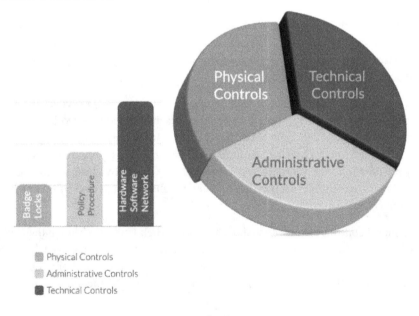

Figure 3.1 – Types of defensive controls

A holistic DiD strategy aims to harmonize the various control types we've explored. Initiating from the physical perimeter defense, it's critical to enforce robust administrative protocols that empower every team member to implement effective technical security controls.

Now that we've outlined the diverse categories of security controls, the focus shifts to perceiving them through a layered lens. This layered approach strengthens an organization's resilience against evolving threats, amalgamating physical, administrative, and technical measures to create a formidable defense posture.

Security layers – Protecting perimeters to information

In the understanding of the dynamics of secure systems, the lens through which we view an organization's defenses plays a crucial role. This section delves into the multifaceted perspective required to encompass the diverse layers of security measures, spanning from the outer perimeters to the core data protection strategies. Each layer presents a unique viewpoint, emphasizing the significance of adopting a comprehensive approach that spans the entire spectrum of an organization's security landscape. By scrutinizing the system through distinct lenses, from the outermost boundaries guarding against external threats to the inner sanctum securing sensitive data, we uncover the necessity of layered security as a means of covering extensive ground in safeguarding an organization's assets and integrity.

This shift is crucial, for it acknowledges that security cannot be viewed through a single, narrow lens. Focusing solely on the perimeter, for instance, might leave internal vulnerabilities lurking. Conversely, a data-centric viewpoint, while vital, risks neglecting the critical first lines of defense at the network edge. By adopting a layered perspective, we create a comprehensive security strategy, attending to each crucial level – from the outer bastion of perimeter defenses to the innermost sanctum of data itself. This holistic approach ensures no chink remains unarmored and no pathway is left unguarded. In this section, we'll delve deeper into this layered paradigm, dissecting the specific security considerations at each level and showcasing how their coordinated interplay builds an impenetrable shield against contemporary threats.

Perimeter defense

Perimeter defense stands as the initial line of defense fortifying an organization's digital boundaries. This layer forms the protective barrier between the external environment and the internal network, acting as a gatekeeper to filter incoming traffic. It encompasses an array of controls and technologies designed to scrutinize and regulate the ingress and egress of data, aiming to detect and repel potential threats before they breach the network.

Deploying firewalls, IDSs, and VPNs serve as primary components of perimeter defense, working collectively to erect an invisible yet formidable shield. While historically fundamental, the evolving threat landscape has expanded the definition of the perimeter, with cloud computing, mobile devices, and remote work necessitating a more dynamic and distributed approach to perimeter security. As such, contemporary perimeter defense strategies demand adaptability and intelligence to counter the increasingly sophisticated threats targeting organizational borders.

Perimeter defense extends beyond technological tools. Security awareness training empowers employees to identify and report suspicious activity and act as human sensors at the network gateway. Vulnerability management practices ensure timely patching of known software flaws, closing potential backdoors before attackers can exploit them. Remember, strong perimeter defense is not just about erecting digital walls; it's about nurturing a vigilant security culture and employing a layered approach that leaves no entry point unchallenged.

Host protection

Once we've secured the outer gates, our journey through the layered defenses descends into the heart of the network: host protection. Here, our focus shifts from filtering network traffic to safeguarding individual devices – the desktops, servers, and mobile endpoints that form the lifeblood of your IT infrastructure. This is where sensitive data is accessed, where applications execute, and where vulnerabilities can become open wounds for attackers to exploit.

Host protection demands a holistic approach, weaving a protective shield around each device. Endpoint security solutions act as vigilant sentries (EDRs), constantly scanning for malware, suspicious activity, and unauthorized applications. Application allowlisting restricts execution only to pre-approved software, limiting the attack surface and preventing malicious programs from gaining a foothold. **DLP** tools monitor sensitive information, preventing its unauthorized transfer or exfiltration, regardless of the device or user involved.

Like what we have seen before, technology alone isn't enough. Patch management practices ensure timely updates and vulnerability fixes, closing potential entry points before attackers can exploit them. User access control restricts privileges based on the principle of least privilege, minimizing the damage one compromised account can inflict. Audit adds another layer to augment the capability of investigation when an incident occurs. By combining robust technical tools with strong security practices, we build a resilient layer of host protection that shields sensitive data and minimizes the impact of potential intrusions.

Remember, the evolution of remote work and the influx of diverse endpoints have amplified the significance of host protection, emphasizing the need for comprehensive defense measures that extend beyond traditional antivirus solutions to encompass advanced threat detection and responsive capabilities.

Operating system and software hardening

Having fortified the perimeter and devices connected to the network, we delve deeper into the very bones of our digital systems. Here, we encounter **operating systems (OSs)** and software running on them, an essential practice focused on minimizing vulnerabilities within the core software components that run our devices. Imagine it as tightening the bolts and welding shut cracks with the aim of minimizing the attack surface, limiting the avenues through which potential threats can infiltrate and compromise the network.

Systems hardening is a two-pronged approach. OS hardening focuses on configuring operating systems to run with the least number of unnecessary features and services. This includes disabling non-essential applications, removing unused accounts and privileges, and applying strict access control measures. This streamlined configuration reduces the potential attack surface and limits the avenues for exploitation.

Software hardening delves deeper into individual applications, applying specific measures to enhance their security posture. This may involve disabling debugging modes, patching vulnerabilities, and implementing security features such as data encryption and code signing. Software hardening involves securing the process from writing code to delivering the application verifiably. By hardening each individual software component, we create a chain of resilient systems, where a breach in one doesn't automatically compromise the entire ecosystem.

> **Minimization is the key**
>
> Software should do exactly what it is supposed to do, nothing more. This is the fundamental idea behind hardening any system: minimization. If you don't need all the packages and libraries that come with the default installation, get rid of them. With every program removed from your OS or application, you are reducing your attack surface.

Remember, OS and software hardening is an ongoing process, demanding constant vigilance and adaptation to evolving threats. Automated tools can assist in configuration management and vulnerability scanning, but human expertise remains crucial in analyzing risks, tailoring configurations, and making informed security decisions. By prioritizing this proactive approach, we ensure our systems remain robust and resilient against digital assaults, minimizing the attack surface and reducing the potential for compromise.

Data protection

Data security stands at the epicenter of safeguarding sensitive information within an organization's ecosystem. This layer focuses on implementing measures to protect the confidentiality, integrity, and availability of critical data assets. The emphasis here revolves around employing encryption, access controls, and robust backup systems to shield against unauthorized access, data breaches, or loss.

Protecting data demands a comprehensive approach, employing a fortress of tools and practices to safeguard confidentiality, integrity, and availability. Data encryption acts as a crucial shield, rendering sensitive information indecipherable to unauthorized entities even if breached, thereby preserving confidentiality. Furthermore, robust access controls and digital signatures ensure that only authorized personnel have appropriate permissions to access specific datasets, thereby safeguarding data integrity and preventing inadvertent or malicious alterations. Alongside these measures, comprehensive backup systems provide an added layer of resilience, ensuring data availability and swift recovery in the event of any disruption or loss.

Chapter 1 emphasized the criticality of comprehending the value associated with the data managed by your organization. Data classification methodologies categorize information based on its sensitivity, ensuring that precise protective measures are employed where they hold the utmost significance. With the increasing prominence of quantum computation, a discussion on forward secrecy becomes imperative. It's essential to acknowledge that technology, by itself, may not suffice; astute decision-making plays a pivotal role. Consider the potential ramifications if your encrypted sensitive data were to be decrypted a decade from now. Therefore, the choice of appropriate protocols remains paramount in ensuring comprehensive security measures.

Remember, data security is not static; it's a continuous journey of vigilance and adaptation. Regular backups and disaster recovery plans ensure data remains accessible, even in the face of unforeseen events. Vulnerability assessments and security audits identify and address weaknesses in data protection measures. By building a culture of data security awareness and constantly refining your safeguards, you ensure your valuable information remains secure and resilient against ever-evolving threats.

Putting it all together, let's visualize how controls from each of these layers help build a more resilient defensive strategy:

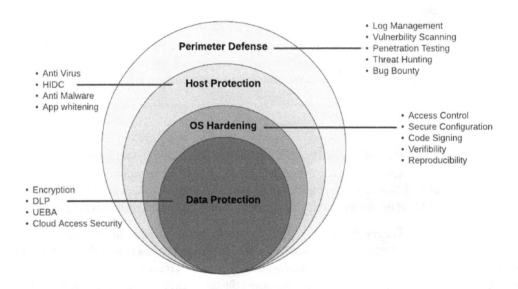

Figure 3.2 – Layers of layered security

The core idea behind adopting a DiD approach toward security revolves around examining the system through various perspectives, ranging from the outermost boundaries defending against external threats to the inner sanctum securing sensitive data, through which we reveal the importance of layered security. This approach efficiently covers a broad spectrum in safeguarding an organization's assets and integrity.

Continuous optimization and adaptation

In the dynamic landscape of cybersecurity, the journey toward establishing a robust security framework does not conclude with the implementation of controls. Instead, it thrives on continuous optimization and adaptive measures. While meticulously constructing layered defenses is crucial, we must recognize that threats evolve, vulnerabilities emerge, and new technologies arise. To maintain a truly resilient posture, we must embrace continuous optimization and adaptation as guiding principles. In this final sub-section, we embark on a critical phase – exploring the ongoing process of assessing, enhancing, and adjusting our security measures to ensure they remain effective, even in the face of a dynamic threat landscape. Remember, building a fortress is only the first step – it's the vigilance and continuous maintenance that ultimately determines its impregnability.

Feedback loops for resilience

Even the most meticulously constructed security framework can develop blind spots over time. That's where feedback loops become vital tools for ensuring continuous improvement and resilience. These feedback loops are iterative processes that capture information from various sources – such as security tools, incident reports, and employee feedback – and feed it back into the security decision-making process. This allows us to identify vulnerabilities, assess the effectiveness of existing controls, and adapt our defenses to evolving threats.

Effective feedback loops require a collaborative approach. Security can't be a solitary pursuit; it thrives on teaming up and fostering a culture of shared responsibility. Each member of the organization, from the CEO to the intern, plays a crucial role in maintaining a secure environment. Encouraging open communication, reporting suspicious activity, and participating in security awareness training are just some ways individuals can contribute to the collective security effort. By breaking down silos and fostering a collaborative mindset, we create a network of interconnected eyes and ears, significantly increasing the overall vigilance and responsiveness of our security posture.

Remember, security is not a race to the finish line; it's a continuous relay where each member of the team plays a critical role. By embracing feedback loops and fostering a culture of shared responsibility, we create a dynamic and resilient security framework that can adapt to the ever-changing threat landscape and safeguard our valuable assets.

Integration and interoperability

In the contemporary threat landscape, siloed security tools operating in isolation provide an incomplete picture. To truly achieve a holistic defense, we must strive for integration and interoperability – forging connections and fostering communication between our diverse security solutions. This allows them to share vital data, automate incident response, and orchestrate a unified response against complex threats. Imagine a scenario where an endpoint detection system identifies a suspicious process. With seamless integration, it can instantly trigger notifications to the firewall, network scanning tools, and incident response teams, enabling a coordinated response to contain and neutralize the threat before it escalates.

As organizations expand, standardizing security protocols becomes pivotal. While individual teams may oversee specific application servers within their scope, security incidents frequently transcend team boundaries. Adhering to a unified standard across teams facilitates shared insights and significantly bolsters security measures by consolidating a wealth of data points, thereby fostering more comprehensive and effective security solutions.

Achieving true integration and interoperability demands careful planning and consideration. Defining standardized data formats, implementing open APIs, and ensuring compatibility across security tools are crucial steps. However, the benefits are substantial. Reduced time to detection and response, streamlined security workflows, and a more comprehensive understanding of your security posture are just some of the rewards reaped from a well-integrated security ecosystem. By eliminating communication barriers and enabling collaboration between security solutions, we create a unified front against increasingly sophisticated threats.

Now let's discuss how to streamline communication when it reaches beyond functional boundaries.

Stakeholder engagement and awareness

The success of security measures depends on the commitment and understanding of every individual within the organization, from the boardroom to the frontline. This necessitates fostering a culture of security consciousness, where everyone takes ownership of protecting sensitive information and upholding security best practices.

Effective stakeholder management begins with understanding individual roles and responsibilities. Tailored security training, incident response procedures, and communication channels should be implemented for each stakeholder group, empowering them to contribute to the collective security effort. Engaging leadership, ensuring transparent communication, and addressing security concerns with clarity and empathy are all crucial to building trust and buy-in. Tapping into an organization's core values and its promises to users about protecting their vital information often serves as a great motivator. Remember, security is not an isolated island; it thrives when woven into the fabric of daily operations and supported by the entire organization.

By nurturing a culture of security awareness and cultivating a proactive mindset among stakeholders, we transform security from a technical barrier into a shared responsibility. This empowered ecosystem, where awareness fuels engagement and ownership, forms the bedrock of a truly resilient security posture, capable of withstanding even the most sophisticated threats.

Now let's turn to how consistent security policies can have exponential effects on an organization's security posture.

Consistency and standardization by security policies

In our earlier exploration of layered security defenses, we laid the groundwork for robust protection. Yet, even the most sophisticated tools require a guiding hand – a clear roadmap for their operation and utilization. In this section, we navigate the intricacies of security policies. Here, we shift focus from individual controls to the overarching framework that governs their application and ensures consistency across the organization.

Imagine a security orchestra, where each instrument – our diverse controls – plays a vital role in the symphony of defense. Security policies act as the conductor, meticulously outlining the score, defining expected behaviors, and ensuring each performer contributes harmoniously to the overall security posture. In this section, we will explore the complete life cycle of security policies, from creating the most effective policies to managing them in a centralized place. By diving into the world of policy, we equip you with the tools to orchestrate a unified defense, maximizing the effectiveness of your existing controls and safeguarding your valuable assets against an ever-evolving threat landscape.

Crafting effective security policies

While security policies might seem like dry documentation, crafting them effectively is an art form crucial to orchestrating a truly resilient security posture. Here, we unpack some key guiding principles to ensure your policies stand as pillars of robust security, not bureaucratic roadblocks.

Clarity and specificity

Policies need to be lucidly articulated, avoiding ambiguity or room for misinterpretation. Every policy should be written in plain language and understandable to all stakeholders, regardless of their technical expertise. Ambiguity breeds confusion, potentially leading to non-compliance and vulnerabilities. Remember, your policies are roadmaps, not riddles. Strive for simple, direct language that clearly outlines expectations and best practices.

Clarity ensures that stakeholders comprehend the expectations and prescribed actions without confusion. Specificity, on the other hand, tailors policies to the unique organizational context, avoiding generic or vague directives that may not be applicable or actionable.

Focus on outcomes, not micromanagement

Instead of dictating every minute action, effective policies focus on defining desired outcomes. This empowers individuals to make informed decisions within the boundaries of the policy, fostering a culture of ownership and engagement. Micromanagement stifles creativity and initiative, while well-defined outcomes guide behavior without stifling critical thinking.

Relevance and adaptability

The threat landscape is dynamic, and your policies must be too. They should possess the flexibility to adapt to emerging risks, technologies, and operational models while remaining relevant and effective. A static policy can quickly become outdated and ineffective. Additionally, policies should be adaptive to varying business requirements and operational nuances, avoiding a one-size-fits-all approach. This adaptability ensures that policies remain pertinent, adjusting to the dynamic nature of security challenges and the organizational landscape.

Risk-based and contextual

Security policies are not one-size-fits-all solutions. They must be tailored to address the specific risks and vulnerabilities unique to your organization and its data assets. Analyze your threat landscape, prioritize risks, and craft policies that directly address your most pressing concerns. A generic policy might offer a false sense of security, while a contextually relevant one tackles your specific vulnerabilities head-on. We will explore how an organization's risk portfolio contributes the most to building its security policies in the next section.

The rationale behind these principles is grounded in the need for policies to resonate with the workforce and operational realities. Clarity and specificity mitigate the risk of misinterpretation or inconsistent adherence to policy directives. Furthermore, relevance and adaptability acknowledge the ever-changing nature of security threats, technologies, and business operations. By being adaptable and tailored, policies remain responsive to the organization's needs, fostering a more resilient and aligned security posture.

Remember, effective security policies are less about dictating actions and more about empowering individuals to make informed decisions within a secure framework. These principles set the stage for crafting policies that work, not just exist.

Risk-informed policies

In the vast ocean of security threats, resources must be directed strategically. Risk-based policies are grounded in a comprehensive understanding of the organization's risk landscape, acknowledging that not all risks are equal in terms of impact or likelihood. Therefore, these policies emphasize a prioritization mechanism that addresses high-impact risks while optimizing resource allocation and mitigating efforts. For instance, if an organization does not use third-party software in any production critical path, it might not be prudent to spend too much time creating a third-party security policy.

Crafting risk-informed policies demands a data-driven approach. Thorough vulnerability assessments and threat intelligence gathering paint a clearer picture of the dangers our organization faces. By analyzing the likelihood and potential impact of different threats, we can prioritize policies that address the most critical risks. This data-driven approach ensures we're not simply chasing shadows but investing our resources in tangible defenses against the most imminent threats.

Remember, we are working with finite resources; this prioritization helps us optimize our security posture, maximizing the effectiveness of our controls and protecting our most valuable assets without overspending or neglecting critical vulnerabilities.

Centralized policy management

As your security framework matures and policies accumulate, managing them effectively becomes a daunting task. Scattered documents, inconsistent enforcement, and manual updates can cripple even the most comprehensive policies. This is where centralized policy management emerges as an essential tool, bringing order to the security policy landscape.

Streamlining policy inception and dissemination

Imagine a world where security policies reside in a single, accessible repository, no longer hidden in forgotten folders or tucked away in individual email inboxes. A common platform provides a centralized hub for authoring, reviewing, and approving policies, streamlining the creation process, and ensuring consistency across all versions. This eliminates discrepancies, fosters collaboration, and simplifies distribution, ensuring everyone is operating from the same playbook.

Automation for scalability and flexibility

Policy updates and distribution are no longer a manual marathon. Custom tools can leverage automation to handle these tasks effortlessly. Automatic notifications alert stakeholders to new and updated policies, while integration with other security solutions ensures seamless implementation and enforcement. This reduces human error, minimizes operational overhead, and improves the overall agility of your security posture.

Secondly, centralized policy management nurtures scalability and flexibility within the organization's security infrastructure. It provides a centralized repository for policies, allowing for swift scalability as the organization expands or adapts to changing security needs. This flexibility enables a more agile response to emerging threats and regulatory changes, facilitating quicker adjustments and deployments of updated policies across the organization.

Visibility and enforcement

A centralized policy management system significantly streamlines the monitoring and enforcement of security policies. Through centralized visibility and control, security administrators gain a comprehensive overview of policy adherence, exceptions, and potential vulnerabilities. This centralized oversight simplifies the monitoring process, allowing for prompt identification of policy violations or weaknesses, thus enabling timely corrective actions to bolster the organization's security posture.

Now that we've delved into the creation and management of security policies, let's shift our focus to real-world illustrations. These practical scenarios will serve as guiding principles, offering insights into how to establish best practices across your organization.

Streamlining security practices

Unless you're in a company that builds security products, security is not top of mind for everyone. So far, we have explored how crucial security policies become standardizing development practices across a company. In this section, we will explore how you can use security policies as a tool to transfer knowledge and share best practices with the rest of your company. Developers often spend a lot of time optimizing the performance of their applications, while at the same time, they care a lot about the reproducibility and reliability of the software they write. While reproducible code makes it easy to debug, a reliable design makes sure it is hard to introduce bugs to the system. Security is not far from these practices. Security's primary goal might not be building bug-free software, but it shares a lot of concepts with a mature software development framework.

A very common policy that many organizations adopt as part of their software development policies is **code review**. This often necessitates getting your code reviewed by another individual (who may be an owner) to ensure code quality and functionality on the production code base. This is also an essential security policy, as the ability to unilaterally update production code not only increases the quality of the software but also protects it from unwanted modification.

Another common software development best practice is to keep your production environment out of human reach. Imagine you have a new intern joining your team and they accidentally delete your production database through a simple tweak in a configuration. Events such as these are not uncommon in the real world. While this is a reliability issue at first glance, it is a much bigger security issue. All the robust DiD strategy can go to waste when an attacker gets hold of an employee's account and intentionally inserts malware into production without any resistance (see the Twitter hack in *A real-world anecdote* from *Chapter 2*).

These are merely two highlights to exemplify how security practices can have benefits well beyond just securing a piece of software. It is often advisable to streamline your organization's security policies and ingrain them into your **software development life cycle** (**SDLC**). Streamlining security practices through clear policies offers several advantages. It minimizes confusion and ambiguity, ensuring everyone within the SDLC understands their security responsibilities [6]. It helps prioritize security considerations early in the development cycle, preventing costly rework and remediation later. Moreover, it fosters a culture of security awareness and ownership, where everyone plays a role in building and maintaining a secure software ecosystem.

The power and benefits of consistent security policies

Consistency in security policies holds immense power within an organization, serving as the linchpin that establishes uniform standards and practices. It ensures a coherent approach across departments and teams, providing a shared understanding of security requirements and guidelines. Through consistent policies, companies can streamline security measures, fostering predictability and reliability in their defenses. This standardization acts as a compass, guiding employees to navigate security challenges with a unified framework, reducing ambiguities and inconsistencies in handling potential threats.

Imagine a well-oiled machine, each gear working in perfect harmony. Consistent policies act as the blueprint for this machine, ensuring every cog, from the CEO to the intern, operates within the same security framework. This standardization fosters a culture of shared responsibility, where everyone understands their role in protecting sensitive information and upholding security best practices.

Let's look at the benefits of implementing security policies:

- **Reduced risk**: By establishing baseline security controls and procedures, consistent policies minimize the likelihood of human error and inadvertent breaches.

- **Enhanced efficiency**: Streamlined processes and unified expectations eliminate confusion and duplicated efforts, saving time and resources.

- **Improved compliance**: Standardized policies facilitate adherence to industry regulations and internal data security mandates.

- **Faster response**: Clear communication channels and defined incident response procedures enable quicker and more efficient handling of security threats.

- **Greater awareness**: Consistent policies cultivate a security-conscious culture, where everyone actively participates in safeguarding organizational assets.

Consistency is the cornerstone of effective security. By implementing and maintaining well-defined, universally applicable policies, you build a robust and resilient defense against evolving threats. Remember, your policies are not simply documents; they are the lifeblood of your security posture, the unifying force that empowers your organization to navigate the turbulent waters of the digital world with confidence.

Compliance and regulatory requirements

Having delved deeply into the foundational role of security policies within an organization's security program, it's time to explore the complementary role played by regulatory requirements in shaping these policies. Companies today face a multitude of regulatory obligations and standards that dictate how data should be handled, stored, and protected. This section delves into the intricacies of compliance, exploring the dynamic interplay between industry-specific regulations, international standards, and the evolving threat landscape. Understanding and adhering to these mandates isn't just a legal necessity; it's a strategic imperative in safeguarding sensitive information and maintaining organizational credibility.

Within the realm of compliance, organizations are tasked with aligning their security measures with a complex web of regulations, including but not limited to GDPR [7], HIPAA [8], PCI DSS [9], and industry-specific mandates. This section aims to dissect these frameworks, unravel their nuances, and illuminate the critical role they play in shaping security policies. It emphasizes the need for proactive measures that transcend mere checkbox compliance, advocating for a comprehensive approach that integrates regulatory adherence seamlessly into the fabric of an organization's security strategy.

Navigating this landscape demands a clear understanding of the key players and their expectations. Regulatory bodies, such as the GDPR or HIPAA, impose specific data protection and privacy mandates. Industry consortia, such as PCI DSS and NIST Cybersecurity Framework [10], establish best practices for specific sectors. Each framework carries its own weight, impacting your business operations and data governance. Demystifying these requirements and aligning your security posture with them is a crucial step in achieving true data security maturity.

Remember, compliance and regulations are not simply hurdles to overcome. They represent a shared responsibility – a commitment to responsible data stewardship and building trust with your stakeholders. By proactively embracing compliance, you not only mitigate legal risks and penalties but also gain a competitive edge, demonstrating your commitment to ethical data practices and secure operations. This section equips you with the knowledge and strategies to navigate the terrain of compliance, ensuring your organization thrives in a data-driven world where trust and security are paramount.

Understanding the regulatory landscape

The crucial first step you need to take before setting the compliance roadmap for your organization is initiating an exhaustive audit of business processes, data flow, and information repositories within the organization. This includes reviewing how data is collected, processed, stored, and shared across various departments and systems. Additionally, mapping out the geographic scope of operations aids in determining the jurisdictions under which the organization falls.

When facing the vast ocean of regulations, pinpointing those relevant to your organization can feel daunting. However, proactive strategies can streamline this process and ensure you focus on the specific frameworks impacting your operations. Let's explore strategies you can take to kick off the regulatory compliance program at your organization:

- **Conducting a business impact assessment**: Begin by conducting a thorough internal assessment. Analyze your data collection and processing practices, the geographic locations you operate in, and the types of data you handle. This introspection sheds light on potential regulatory touchpoints and narrows down the search criteria.

- **Consulting expert resources**: Leverage readily available resources to guide your journey. Consult regulatory agency websites, industry-specific compliance associations, and legal counsel familiar with your sector. These experts can provide valuable insights into relevant regulations and potential applicability based on your business profile.

- **Embracing regulatory mapping tools**: Consider utilizing technology. Several online tools and platforms offer regulatory mapping services, helping you match your business activities to specific regulatory requirements. These tools can act as efficient filters, saving you time and effort in navigating the initial identification phase.

Much like everything else in security, the identification process is not a one-time venture. Regulatory landscapes evolve, and your business might grow or pivot, influencing the applicable frameworks. Regular review and re-evaluation of your compliance scope are crucial to ensure continued alignment and minimize the risk of non-compliance. A fundamental concept is to continuously consider whether the collection of new data introduces additional regulatory obligations.

By implementing the aforementioned strategies, you transform potential regulatory confusion into a clear roadmap for compliance. This initial clarity lays the foundation for building a robust and targeted compliance strategy, ensuring your organization sails smoothly through the regulatory waters with confidence.

Now that we understand how we can streamline the relevant regulatory requirements that affect us, let's shift gears to understand the underlying principles of some common ones.

Regulatory bodies worldwide impose varying standards, mandating adherence to specific guidelines for data protection, privacy, and overall security. Delving into the world of compliance and regulatory requirements can feel like navigating a complex archipelago, with each regulatory body a distinct island demanding attention. To steer your organizational security posture through this landscape effectively, a clear understanding of the common patterns and specific intricacies is crucial. The overarching patterns among these bodies often revolve around safeguarding sensitive information, ensuring secure storage and transmission, maintaining data integrity, and establishing user privacy rights. Some common themes include the following:

- **Data protection and privacy**: Most regulations mandate robust data protection and privacy measures, encompassing data minimization, access control, data breach notification, and individual rights regarding their data.

- **Security controls**: Implementing appropriate security controls to safeguard sensitive data is a cornerstone of most compliance regimes. This involves leveraging encryption, vulnerability management, incident response plans, and ongoing security assessments.

- **Governance and accountability**: Establishing clear governance structures and accountability mechanisms are vital to demonstrating adherence to regulations. This includes designated data protection officers, data inventories, and risk management processes.

- **Transparency and reporting**: Regularly reporting on compliance efforts and potential data breaches is often a requirement, fostering trust and demonstrating commitment to responsible data handling.

We will now discuss some prominent regulatory frameworks to illustrate the specific expectations and complexities involved.

The General Data Protection Regulation (GDPR)

This EU regulation stands as a prominent example of regulatory frameworks shaping the global data landscape. Applicable to any organization processing the personal data of EU citizens, the GDPR emphasizes users' individual rights and control over their data. Here are some key requirements that businesses must navigate:

- **Transparency and accountability**: Organizations must be transparent about their data practices, provide clear notices and information to individuals, and demonstrate robust data governance structures.

- **Data subject rights**: Individuals have extensive rights regarding their data, including the right to access, rectify, erase, restrict processing, and object to automated decision-making.

- **Lawful basis for processing**: Data processing must be grounded on a legal basis, such as consent, contractual necessity, or legitimate interest. Organizations must clearly articulate and demonstrate the relevant basis for each data processing activity.

- **Security controls**: Implementing appropriate technical and organizational safeguards to protect against unauthorized access, data breaches, and loss of data is paramount.

- **Data breach notification**: In case of a data breach, prompt notification to regulators and potentially affected individuals is mandatory.

Remember, this is just a brief overview. The GDPR encompasses a wide range of detailed provisions, and seeking expert guidance is recommended for ensuring comprehensive compliance.

The Health Insurance Portability and Accountability Act (HIPAA)

In the healthcare realm, HIPAA takes center stage, safeguarding the privacy and security of **protected health information (PHI)**. Applicable to covered entities such as healthcare providers, health plans, and clearing houses, HIPAA sets forth crucial requirements for handling such sensitive data:

- **Minimum necessary standard**: Data collection and utilization must be limited to the minimum necessary for the specific healthcare purpose.

- **Security safeguards**: Implementing appropriate administrative, physical, and technical safeguards to protect PHI from unauthorized access, disclosure, or misuse is mandatory.

- **Patient rights**: Individuals have the right to access, copy, amend, and request restrictions on the use and disclosure of their PHI.

- **Breach notification**: In case of a breach affecting PHI, timely notification to affected individuals and the Department of Health and Human Services is required.

Once again, remember that this is just a snapshot, and seeking guidance from healthcare compliance specialists is recommended for the comprehensive understanding and implementation of HIPAA's extensive provisions.

The Payment Card Industry Data Security Standard (PCI DSS)

This industry-driven standard focuses on safeguarding payment card information. It outlines specific technical controls, vulnerability management practices, and incident response procedures. For organizations processing cardholder data, PCI DSS acts as a vital line of defense. Encompassing a broad range of controls, PCI DSS safeguards sensitive payment information and minimizes the risk of financial breaches. Here are some key requirements that organizations must address:

- **Build and maintain a secure network**: This involves deploying firewalls, encrypting cardholder data, and regularly patching vulnerabilities in systems and applications.

- **Protect cardholder data**: Implementing strong encryption for data at rest and in transit, restricting access to authorized personnel, and employing secure tokenization solutions are crucial aspects.

- **Maintain a vulnerability management program**: Regularly scanning systems for vulnerabilities, promptly patching identified flaws, and updating security software are essential to staying ahead of threats.

- **Implement strong access control measures**: Applying principles of least privilege, multi-factor authentication for sensitive systems, and logging and monitoring user activity are vital to data security.

- **Regularly test and monitor systems**: Conducting penetration testing, validating security controls, and monitoring systems for suspicious activity are key aspects of proactive security.

Again, a mandatory disclaimer: This is a complex technical standard, and seeking guidance from cybersecurity professionals is recommended for ensuring comprehensive adherence.

The Sarbanes-Oxley Act (SOX)

In the realm of financial reporting and internal controls, the **Sarbanes-Oxley Act** (**SOX**) [11] stands tall, holding publicly traded companies accountable for accurate and reliable financial reporting. While not directly focused on data security, SOX necessitates robust **internal control over financial reporting** (**ICFR**) to prevent and detect errors or fraud. Here are some key requirements that companies must embrace:

- **Management's assessment of internal controls**: CEOs and CFOs must annually assess the effectiveness of ICFR, ensuring the adequate design and operation of internal controls over financial reporting processes.

- **Internal control documentation and testing**: Companies must document and maintain comprehensive descriptions of their ICFR structure, followed by ongoing testing and evaluation to ensure its effectiveness.

- **Disclosure of internal control weaknesses**: Any material weaknesses or significant deficiencies in ICFR must be disclosed in financial reports, demonstrating transparency and fostering investor confidence.

- **Compliance with additional standards**: SOX compliance necessitates adherence to specific accounting standards set by the **Public Company Accounting Oversight Board** (**PCAOB**), outlining detailed requirements for financial reporting and auditing practices.

Gentle reminder: SOX compliance goes beyond financial data and impacts various internal processes and controls. Seeking guidance from financial and compliance experts is recommended for a comprehensive understanding and implementation of SOX provisions.

US Executive Order 14028 (SBOM)

A more recent one, Executive Order 14028 [12], shines a spotlight on bolstering the defenses of **critical infrastructure entities** (**CIEs**) against cyber threats. Applicable to a wide range of sectors, from energy and finance to communication and healthcare, the order outlines various requirements to enhance cyber resilience:

- **Incident response and reporting**: CIEs must develop and implement rapid and effective incident response plans, including timely reporting of cyber incidents to the **Cybersecurity and Infrastructure Security Agency** (**CISA**).

- **Multi-factor authentication** (**MFA**): Implementing MFA for privileged access and remote access to critical systems becomes mandatory, significantly reducing the risk associated with compromised credentials.

- **Threat modeling and risk management**: Conducting regular threat modeling exercises and risk assessments helps CIEs identify vulnerabilities and prioritize mitigation efforts, enabling proactive defense strategies.

- **Security logging and monitoring**: Continuous logging and monitoring of critical systems and sensitive data is crucial to early detection of suspicious activity and potential intrusion attempts.

- **Secure software development practices**: Integrating secure coding practices, vulnerability scanning, and supply chain security throughout the software development life cycle strengthens the cyber posture of deployed systems.

Navigating these core requirements is vital for CIEs to fulfill their responsibility to national security and public safety. Remember, EO 14028 represents a continuous commitment to improving cybersecurity across critical infrastructure, demanding ongoing assessment, mitigation, and evolution of existing security practices.

In this ever-evolving regulatory landscape, these examples merely scratch the surface. They underscore the commonalities shared across various compliance frameworks. More significantly, safeguarding the data entrusted to us by customers shouldn't solely hinge on regulations to ensure its integrity. It's a rallying call for our industry to elevate standards proactively, and to set a higher benchmark for data protection and privacy.

Regulatory compliance isn't a fixed endpoint but an ongoing voyage. Cultivating a culture of awareness, actively tracking regulatory changes, and customizing your security approach to align with specific frameworks empower you to navigate compliance complexities confidently. This approach minimizes risks and ensures your organization adheres to the dynamic legal and ethical boundaries of the digital landscape.

Aligning security with regulations

Navigating the complex landscape of regulations and compliance requirements can feel like chasing a moving target. Often, organizations fall into the trap of implementing security measures solely to tick compliance boxes, overlooking the deeper value proposition of a truly secure environment. However, a far more effective and sustainable approach lies in shifting the mindset from regulations as drivers to data privacy as the fundamental principle.

Instead of viewing compliance as a separate entity dictating security, the emphasis should shift toward an intrinsic incorporation of user data protection. By adopting a proactive approach centered on safeguarding user information as an inherent organizational responsibility, companies can transcend the limitations of mere compliance adherence.

Imagine constructing your security posture not with compliance mandates as individual bricks, but with user data protection as the solid foundation. This data-centric approach transcends the limitations of specific regulations, recognizing that protecting personal information is not just a legal obligation, but an ethical imperative and a strategic advantage. This shift in perspective unlocks several key benefits:

- **Unified and efficient security framework**: By focusing on the common denominators of various regulations – data privacy, access control, and breach notification – you not only build a holistic security framework that transcends specific compliance mandates, but you also start to understand their underlying essence. This eliminates redundancy and streamlines security efforts, resulting in a more efficient and cost-effective approach.

- **Enhanced security posture**: User data protection, by its very nature, demands robust security controls. Investing in encryption, robust access management, and vulnerability management not only addresses compliance mandates but also elevates your overall security posture, mitigating a wider range of threats and protecting against evolving cyberattacks.

- **Building trust and transparency**: Aligning your security strategy with data privacy fosters trust and transparency with your users and stakeholders. Demonstrating a commitment to protecting their information, proactively communicating your practices, and empowering them with control over their data builds deeper relationships and strengthens your brand reputation.

- **Future-proofing your security program**: The regulatory landscape is constantly evolving. By anchoring your security program in data privacy principles, you create a flexible and adaptable foundation that can easily adapt to new regulations and emerging threats. This ensures your security posture remains relevant and effective in the long run.

Shifting from compliance-driven to data-centric security does not imply disregarding regulations. On the contrary, it strengthens your adherence by creating a more robust and enduring security framework. By prioritizing user data protection as the cornerstone of your security strategy, you not only fulfill legal obligations but also build a foundation for a future-proofed, adaptable, and resilient organization. Remember, true security is not about checking boxes; it's about building trust, safeguarding sensitive information, and empowering your users – a paradigm shift that elevates your organization to a higher plane of data stewardship and operational excellence.

Now let's explore how compliance can benefit security teams driving for consistency across an organization.

Compliance as a catalyst for consistency

For security teams often battling against budget constraints and fragmented responsibilities, compliance frameworks can feel like rigid hindrances, forcing adherence over innovation. However, a closer look reveals that compliance holds an underappreciated potential: acting as a powerful catalyst for consistency, driving the implementation and enforcement of crucial security practices across an entire organization.

Imagine a sprawling organization, its security posture resembling a patchwork quilt sewn from disparate initiatives and individual team efforts. Regulatory mandates, in this scenario, become unifying threads, weaving consistency into the fabric of your security program. They provide a shared language – a set of baseline expectations that apply to all departments, from engineering to marketing to finance. This standardization translates into several tangible benefits:

- **Reduced risk and improved security posture**: As discussed repeatedly, compliance requirements often mandate the implementation of critical security controls, such as vulnerability management, access control, and incident response plans. By embracing these requirements, you elevate your overall security posture across the organization, minimizing vulnerabilities and strengthening defenses against a wider range of threats.

- **Streamlined operations and cost efficiency**: Regulatory frameworks act as guides, streamlining security processes and ensuring consistent application across the organization. This avoids duplication of effort, reduces the need for redundant tools, and ultimately fosters operational efficiency and cost savings.

- **Enhanced collaboration and communication**: Compliance mandates breaking down siloed walls, forcing collaboration between teams and departments on critical security initiatives. This fosters clearer communication, shared responsibility, and a unified approach to data protection.

- **Improved visibility and reporting**: Many regulations emphasize robust reporting and record-keeping practices. These requirements provide valuable insights into your security posture, enabling data-driven decision-making and proactive identification of potential weaknesses.

By leveraging compliance as a catalyst for consistency, security teams transform a perceived burden into a strategic advantage. They establish a common ground, unite diverse functional teams around shared goals, and elevate the overall security posture of the organization. Remember, compliance is not just a checkbox exercise; it's a foundation for building a resilient, agile, and cost-effective security program that safeguards your organization against evolving threats and protects your most valuable asset – your data.

Let's now turn our attention to some tricks to make compliance work for us.

Compliance as a tool, not a goal

The enticement of reaching the compliance "finish line" can be seductive. Ticking off boxes, achieving certifications, and basking in the warm glow of regulatory approval feels like a victory. However, a crucial distinction must be made: compliance is not a destination but a tool on the journey toward robust security. Viewing it as an endpoint risks complacency and a false sense of security, often called **security theater**.

> **Note**
> Security theater in the tech industry refers to measures designed to make people feel more secure without improving security. These actions often look impressive and create an illusion of safety, but they don't effectively address real security threats.

True security demands a continuous ascent – an ongoing process of risk assessment, vulnerability mitigation, and proactive adaptation. Here's why viewing compliance as a tool empowers this climb:

- **Flexible framework for adaptation**: Regulations, while evolving, offer a structured framework that can be adapted to your specific industry, threat landscape, and organizational needs. This flexibility allows you to build upon compliance requirements, layering additional controls and proactive measures that address unique vulnerabilities and emerging threats.

- **Proactive risk management**: Compliance mandates often act as a starting point for risk assessments, helping you identify and prioritize security weaknesses beyond regulatory requirements. This proactive approach allows you to address critical vulnerabilities before they become exploitable, minimizing potential damage and reputational harm.

- **Culture of continuous improvement**: Viewing compliance as a tool fosters a culture of continuous improvement within your organization. Teams focus on optimizing security practices, exceeding compliance baselines, and proactively seeking out best practices in data protection and threat mitigation. This ongoing momentum keeps your security posture agile and resilient in the face of constant change.

Remember, compliance is a valuable tool – a roadmap on your security journey. But it's not the destination. By leveraging its framework, embracing its flexibility, and using it as a springboard for ongoing improvement, you can transform compliance from a checkbox exercise into a driving force for a secure, adaptable, and future-proof organization.

Measuring compliance effectiveness

Having embraced compliance as a powerful tool for driving consistent and proactive security, the next crucial step is measuring its effectiveness. Just as a navigator relies on precise measurements to ensure a safe course, security teams need an accurate understanding of how effectively their compliance efforts translate into a real-world security posture. This requires moving beyond simply ticking boxes and delving into a more nuanced assessment.

Evaluating compliance effectiveness involves focusing on key metrics beyond mere regulatory adherence. These can include the following:

- **Vulnerability management**: Track the rate of vulnerability identification, patching, and remediation to assess how effectively compliance requirements translate into tangible improvements in your security posture.

- **Incident response performance**: Measure the timeliness and effectiveness of incident response procedures, analyzing how compliance frameworks contribute to faster detection, containment, and recovery from security incidents.

- **User security behavior**: Gauge the success of compliance-driven security awareness training by monitoring user adherence to secure password practices, reporting of suspicious activity, and overall understanding of data protection policies.

- **Internal audits and penetration testing**: Regularly conduct independent assessments to identify potential gaps between the compliance framework and its actual implementation within your organization. These audits and penetration tests offer valuable insights into areas requiring improvement and provide a more holistic picture of compliance effectiveness.

By analyzing these metrics and continuously refining your compliance program, you ensure that your organization remains on a secure course, navigating the ever-evolving landscape of regulations and threats with confidence. As we discussed numerously before, measuring compliance effectiveness is not a one-time exercise; it's an ongoing process, and a feedback loop that informs your security strategy and ensures your compliance investments translate into a robust and resilient security posture.

Enforcement and accountability

After punctiliously assembling the layers of our security fortress, from layered defenses to consistent policies and compliance strategies, we arrive at a crucial final piece of the puzzle: enforcement and accountability. Imagine building an intricately designed vault, only to leave the keys on the doorstep. Effective security requires not just robust controls but also the mechanisms to ensure their rigorous application and hold individuals accountable for maintaining our security posture.

In essence, enforcement and accountability form the backbone of a security culture that emphasizes the practical application of security measures. These elements create a framework where security isn't just a set of rules, but a collective responsibility ingrained within the organization's DNA. By setting clear expectations, monitoring compliance, and establishing consequences for non-compliance, organizations establish a proactive approach to security governance.

This section delves into the practical realities of enforcement and accountability in the context of our DiD strategy. We explore the strategies, processes, and technologies that empower organizations to operationalize their security policies, ensuring consistent adherence at all levels. We'll examine strategies for fostering a culture of security awareness and responsibility, where individuals understand their role in protecting sensitive information and face clear consequences for non-compliance.

Validation and assurance

Ensuring the effectiveness of security controls within a system extends far beyond their initial design. Consider a scenario where the implementation of a security measure, such as **mutual TLS (mTLS)** for internal server communication, was pursued without adequate infrastructure to manage the associated certificates. In this instance, despite the perceived enhancement to the organization's security posture, a critical oversight occurred in establishing a robust public key infrastructure. Consequently, the deployment utilized insecure self-signed certification, ultimately exposing vulnerabilities. Subsequently, an attacker exploited this vulnerability by inserting an unauthorized service into the communication flow, resulting in the unauthorized extraction of extensive user data several months later.

With our security framework meticulously constructed, a nagging question inevitably arises: are our defenses truly effective? This is where validation and assurance play a crucial role, where we rigorously test and measure the resilience of our carefully erected safeguards. This vital process goes beyond mere compliance checks, delving into the practical effectiveness of our security controls in the face of real-world threats.

Several key techniques and strategies empower us to achieve this assurance:

- **Security assessments and audits**: Independent assessments by skilled professionals provide an objective evaluation of our security posture, identifying potential vulnerabilities and weaknesses in our controls. Penetration testing simulates real-world attack scenarios, exposing exploitable loopholes and validating the effectiveness of our response capabilities. This can be led by either an internal group of penetration testers or a third party for simulating external attacks.

- **Red teaming and blue teaming exercises**: These simulated adversarial engagements pit dedicated "red teams" against our "blue team" defenders. This controlled conflict environment allows us to test our incident response procedures, communication channels, and decision-making skills under pressure, revealing areas for improvement before facing an actual attack.

- **Continuous monitoring and logging**: Utilizing advanced monitoring tools and log analysis platforms equips us with real-time visibility into system activity, user behavior, and potential anomalies. This continuous vigilance enables early detection of suspicious activity, potential breaches, and deviations from established security baselines.

- **Vulnerability management and patching**: Proactive vulnerability management practices, including regular scans, prioritization, and timely patching, ensure our defenses remain up-to-date and effectively counter known exploits. This meticulous approach mitigates the risk of attackers capitalizing on unpatched vulnerabilities in our systems.

Absolute certainty in the face of constant evolution and unknown threats is elusive, but confidence in our security posture is attainable. By proactively employing a multifaceted approach combining independent assessments, simulated attacks, continuous monitoring, and vigilant vulnerability management, we gain valuable insights into the real-world effectiveness of our security controls. These continuous validation exercises illuminate potential weaknesses, guide our strategy adjustments, and fuel a culture of improvement. While no single technique offers foolproof assurance, this ongoing cycle of testing, analysis, and adaptation allows us to refine our defenses, mitigate risks, and face the ever-shifting threat landscape with a well-founded sense of preparedness. Remember, security is a journey, not a destination, and embracing the constant quest for validation ensures our controls remain fit for purpose, safeguarding our precious data and our organizational resilience.

Let's now switch gears. We've established a crucial foundation: the bedrock principle that security demands vigilance and responsibility throughout the entire organization. While this shared ownership forms the core of our security posture, it begs the question – how do we translate this principle into tangible action? Let's dive deeper into the practical realm, exploring a toolbox of strategies to empower individuals and build a robust culture of security excellence.

Shift-left security

In the quest for robust security, a paradigm shift is gaining traction: **shifting security left**. Departing from the traditional model of placing security checks primarily at the end of the development life cycle, this approach embeds security considerations throughout the entire software development process. Imagine crafting a tapestry, meticulously weaving threads of security best practices into the very fabric of the application from its inception.

Shifting left offers several compelling advantages:

- **Early detection and remediation**: Integrating security testing and vulnerability assessments into early development phases allows for the identification and patching of flaws much earlier in the cycle. This significantly reduces the cost and complexity of fixing issues compared to addressing them after deployment.

- **Proactive design and architecture**: Embracing security by design principles encourages developers to consider security implications alongside functionality from the outset. This leads to the creation of inherently secure applications, minimizing the vulnerabilities that attackers can exploit.

- **Improved collaboration and ownership**: Shifting left fosters closer collaboration between developers and security teams, breaking down silos and promoting a shared responsibility for the application's security posture. This collaboration facilitates knowledge sharing and builds a culture of security awareness throughout the organization.

- **Continuous improvement**: Embedding security practices within the development workflow enables iterative testing and feedback loops. This ongoing process continually strengthens the application's security posture, adapting to evolving threats and vulnerabilities in real time.

However, shifting left comes with its own challenges, requiring changes in tools, processes, and organizational culture. Implementing DevSecOps practices, integrating security automation, and educating developers on secure coding principles are crucial to successful implementation.

Now that we understand the need for cross-functional alignment in building a truly resilient defense system, let's explore how we can streamline our communication.

XFN collaboration

It is quite clear that in the intricate strategy of DiD, security extends beyond mere technological fortifications such as firewalls and encryption algorithms; it necessitates cohesive collaboration across various disciplines. This is where **cross-functional (XFN)** collaboration emerges as a crucial element: picture developers and security teams operating like a finely-tuned orchestra, each contributing essential parts to create a harmonious, impenetrable security structure. Robust XFN relationships, founded on trust and shared comprehension, enable this transformative capability.

Now, cultivating this rapport might seem intangible – a soft skill lost in the binary world of code. But consider the developer's perspective: reliable systems, smooth deployments, and minimized downtime are just as alluring as impenetrable defenses. Tap into these shared ambitions! Organize security awareness workshops tailored to developer challenges, highlighting vulnerabilities that can lead to service disruptions and data breaches. Showcase case studies where proactive security measures prevented costly outages, saved development hours, and boosted user trust. By framing security as a partner in achieving common goals, you bridge the divide, fostering collaboration and weaving XFN relationships into the very fabric of your security posture.

Evolving security responsibilities

In the rapidly evolving landscape of cybersecurity, the traditional approach of confining security responsibilities within a dedicated team no longer aligns with modern organizational structures or security demands. The once isolated role of security practitioners within a team of 20 for a company of 1,000 has become increasingly inadequate. Today's security challenges necessitate a shift toward a collective responsibility model where security is embraced by everyone within the organization, irrespective of their specific roles.

Imagine a sprawling cityscape: a thousand bustling buildings, interconnected infrastructure, and a vibrant pulse of activity. Can you envision securing this metropolis with just 20 dedicated guards patrolling its perimeters? Of course not. Modern security demands a distributed network of vigilance, where responsibility extends beyond a select team and permeates every corner of the organization. This, my friends, is the call to action for **evolving security responsibilities**.

Today's dynamic landscapes, with cloud-based services, distributed applications, and a workforce spanning continents, necessitate a paradigm shift. We need to embrace a shared ownership model, where every individual – from developers crafting code to executives charting strategy – actively participates in safeguarding our digital ecosystems.

This evolution signifies a pivotal paradigm shift – a transition from perceiving security as a compartmentalized task to acknowledging it as a shared and integrated responsibility across the entire company. Each individual, from software developers and system administrators to marketing professionals and top-level executives, must be equipped with a fundamental understanding of security principles and practices. Embracing this collective responsibility approach empowers every member of the organization to contribute to a secure environment, actively participating in safeguarding data, systems, and infrastructure against evolving threats.

Think of it as a decentralized fire alarm system. Every smoke detector and every sprinkler head plays a crucial role in preventing and mitigating disasters. Similarly, empowering individuals with security awareness training, access controls, and incident reporting tools equips them to become active nodes in our security network. This distributed vigilance not only multiplies our defensive power but also fosters a culture of proactive responsibility, where securing our systems becomes everyone's business.

Remember, this shift isn't a threat, but an opportunity. By empowering individuals and fostering XFN collaboration, we build a resilient security posture, capable of adapting to evolving threats and safeguarding our organization with the collective strength of 1,000 vigilant eyes and 10,000 proactive hands. Embrace the call to action – let's evolve our security responsibilities and build a future where everyone holds the key to a secure and resilient digital realm.

Summary

In this chapter, we revisited many concepts introduced in earlier chapters and built a framework to thoughtfully design a DiD security strategy. We organized our defense mechanism to provide multi-dimensional benefits. As we reiterated, the fundamental challenge of defense is that it must be resilient to attacks every single time. This makes completeness a critical criterion for a robust security strategy. Understanding the fundamental methodologies to divide your defenses into layers will pay dividends throughout your security career. We also delved into standardizing security practices to achieve optimal efficiency. A simple system is almost certainly more secure than a complex one; similarly, a uniform security practice is often more comprehensive than a divided one. We learned how we can use security practices as a tool to drive uniform security posture at a large scale. Another key piece in the puzzle is regulatory compliance. We took a high-level tour of common regulatory requirements and demystified them by finding common themes. This is crucial as we explore how we can utilize compliance as a tool to drive better security practices rather than making it our destination. Finally, we covered the irreplaceable piece of ensuring our defense mechanisms are not sitting idle but actively throttling attackers.

The concepts covered in this chapter should give you a framework for designing a multi-layered defense strategy that can keep the most persistent attackers at a distance. Keep in mind that refining and adjusting your defense is the key to DiD.

Key takeaways

- The security paradigm of "defense in depth" is a journey without a goal. Continuous adaptation and improvement are pivotal in layered security models.

- A comprehensive DiD strategy puts equal emphasis on people, processes, and technology.

- The most secure system in the world is the one that has no valuable assets. The key to maximizing security is minimization, which means deleting more lines of code than you write. This really puts the importance on every line of code you write as an additional attack surface that should be optimized and reduced to a minimal shape.

- When designing a security control, remind yourself of the three key properties of a reference monitor: non-bypassable, tamper-proof, and verifiable.

- Software should do exactly what it is supposed to do, nothing more. Making sure a system does not do anything extra is the hardest part of security.

- Security systems that work well together are almost always better than controls deployed in silo.

- Security policies must be clear, actionable, relevant, and unique to each organization.

- Utilize compliance requirements to raise your security baseline, not to check boxes.

- Security might limit the usability of software in some instances, but often, it comes with positive side effects such as reliability, reproducibility, and verifiability. Tap into these additional benefits when collaborating with XFN teams.

Now that we have established a comprehensive framework for DiD, we will zoom out a bit in the next section of the book, where we will put on our adversarial hat to solidify our defense strategies further. Let's start with understanding how attackers think and build this skill of how to think like an attacker.

Congratulations on getting this far! Pat yourself on the back. Good job!

Further reading

To learn more about the topics that were covered in this chapter, take a look at the following resources:

- [1] NIST definition of DiD: `https://csrc.nist.gov/glossary/term/defense_in_depth`
- [2] Constitutional AI from Anthropic: `https://arxiv.org/abs/2212.08073`
- [3] Wikipedia – HSMs: `https://en.wikipedia.org/wiki/Hardware_security_module`
- [4] Wikipedia – TPM: `https://en.wikipedia.org/wiki/Trusted_Platform_Module`
- [5] NIST reference monitor: `https://csrc.nist.gov/glossary/term/reference_monitor`
- [6] Secure Software Development Framework: `https://csrc.nist.gov/projects/ssdf`
- [7] GDPR: `https://gdpr-info.eu/`
- [8] Health and Human Services: `https://www.hhs.gov/hipaa/index.html`
- [9] PCI Standards: `https://www.pcisecuritystandards.org/`
- [10] NIST Cybersecurity Framework: `https://www.nist.gov/cyberframework`
- [11] SOX compliance: `https://en.wikipedia.org/wiki/Sarbanes%E2%80%93Oxley_Act`
- [12] EO 14028: `https://www.whitehouse.gov/briefing-room/presidential-actions/2021/05/12/executive-order-on-improving-the-nations-cybersecurity/`

Part 2:
Building a Layered Security Strategy – Thinking Like an Attacker

In this part, we focus on attackers to understand the core weaknesses in our systems. A resilient system requires thinking outside the box. To outsmart an attacker, we must first understand their motivations and techniques. We'll dissect the attacker's mindset, helping you reverse engineer their moves. You'll learn how to systematically uncover vulnerabilities in your systems through threat modeling and penetration testing. By mapping out potential attack vectors, you'll gain a tactical advantage. This knowledge will empower you to build a proactive, multi-layered defense strategy that makes breaches far more difficult and costly for malicious actors.

This part has the following chapters:

- *Chapter 4, Understanding the Attacker Mindset*
- *Chapter 5, Uncovering Weak Points through an Adversarial Lens*
- *Chapter 6, Mapping Attack Vectors and Gaining an Edge*
- *Chapter 7, Building a Proactive Layered Defense Strategy*

4

Understanding the Attacker Mindset

Now that we have established a structured framework for building a **Defense-in-Depth (DiD)** security strategy, let's shift our focus toward the attacker mindset. A well-fortified system is only as strong as its defenders' understanding of their adversaries. A very common saying in the security profession is "Thinking like an Attacker." This saying isn't merely a cool catchphrase; it's a fundamental principle essential to crafting truly robust defenses. Don't interpret this as an ask for being an attacker when you are not; this will be like asking a drag racer to think like a fighter jet pilot. These are two completely different things. In the world of security, thinking like an attacker merely means brainstorming what could possibly go wrong in the systems you are trying to protect and making informed, risk-based decisions on what to do about it.

This chapter opens the door to dissecting this critical paradigm, delving into the motivations, methodologies, and maneuvers of the very forces we aim to thwart. We'll begin by peeling back the layers of the attacker's psyche and dissecting the diverse motivations that drive cybercrime, from state-sponsored espionage to financially motivated exploits. Understanding these drivers reveals the potential resources and effort they might invest in breaching your defenses, allowing you to prioritize your own countermeasures accordingly.

But motives are just the tip of the iceberg. We'll delve deeper, scrutinizing the vast arsenal of tactics and strategies employed by different threat actors. From lone wolves wielding social engineering scams to sophisticated hacking groups deploying zero-day exploits, you'll gain comprehensive insights into the attacker's playbook. This knowledge is your secret weapon, enabling you to identify vulnerabilities in your defenses before they're exploited.

By the end of this chapter, you'll be equipped with a clear-eyed view of the adversary landscape. You'll understand the motivations, methods, and tools used by various threat actors, enabling you to do the following:

- **Proactively identify and prioritize vulnerabilities**: Predict where attackers might focus their efforts and prioritize security investments accordingly.

- **Craft targeted countermeasures**: Design defenses tailored to specific attacker types and their preferred strategies.

- **Become a vigilant guardian**: Develop a deeper understanding of the evolving threat landscape and anticipate future attack vectors.

This deeper understanding is not about peering into the abyss; it's about harnessing the darkness to illuminate your own defenses. Embrace the attacker mindset and watch your once-porous fortress transform into an impregnable bastion against the ever-evolving cyber adversary.

In this chapter, we're going to cover the following main topics:

- Exploring the attacker's perspective

- Thinking like an attacker – identifying weaknesses

- Understanding **tactics, techniques, and procedures (TTPs)**

- Defensive countermeasures – turning the tables

Let's get started!

Exploring the attacker's perspective

In the intricate landscape of cybersecurity, understanding the attacker's perspective is paramount for designing effective defensive strategies. This section delves into the mind of an attacker, unraveling the methodologies, tools, and motivations that drive their actions. By dissecting the attacker's mindset, we gain insights into their decision-making processes, target selection criteria, and exploitation techniques.

Our journey begins by peering into the psyche of the cybercriminal. We'll analyze the multifaceted motivations that fuel their endeavors, from state-sponsored espionage to financially motivated exploits. Understanding their goals and risk tolerance informs our own prioritization of vulnerabilities and allocation of resources.

Next, we'll dissect the attacker's arsenal, scrutinizing the ever-evolving tools and techniques employed across the attack life cycle. From rudimentary phishing campaigns to sophisticated zero-day exploits, we'll demystify their functionalities and pinpoint their preferred targets. This technical understanding equips us to identify attack vectors and design countermeasures tailored to specific threat profiles.

But motives and tools are just half the story. We'll delve deeper, unraveling the hidden economy of cybercrime. We'll trace the illicit financial flows, from initial attacks to crypto-laundering networks and underground marketplaces. This economic analysis allows us to anticipate attacker behavior and resource allocation, predicting their target selection and optimizing our own defensive investments.

Finally, we'll turn our attention to the apex predators of the cyber jungle: **advanced persistent threats (APTs)**. These highly skilled, state-backed actors operate with unparalleled resources and persistence. By dissecting their unique TTPs, we gain invaluable insights into the cutting edge of cyberwarfare, preparing ourselves for the most sophisticated attacks imaginable.

By the end of this section, you'll be equipped with a deep understanding of the attacker's landscape. You'll grasp their motivations, methods, and financial drivers, empowering you to design defenses that are not merely reactive, but that proactively anticipate and mitigate even the most intricate and evolving cyber threats. This knowledge is your shield, weapon, and ultimate advantage in the digital arms race.

In the mind of a cybercriminal

In previous chapters, we discussed the existence of a diverse landscape of attackers. Each is driven by their own complex motivations, whether they are state-sponsored espionage rings seeking secrets or opportunistic cyber thieves chasing a quick buck. Analyzing these motivations is key. It reveals which adversaries pose the greatest threat to your organization, where they'll focus their efforts, and the resources they'll bring to bear.

In 2017, Equifax's security posture, tailored to financially motivated hackers, proved insufficient when faced with a state-sponsored espionage ring seeking sensitive consumer data. The attackers, motivated by geopolitical goals, leveraged a known vulnerability in a piece of third-party software, not initially prioritized due to its perceived low financial risk. This breach highlights the value of understanding diverse attacker motivations. Had Equifax broadened its threat landscape beyond financial attackers, it might have prioritized patching that specific vulnerability, potentially preventing the massive data exfiltration that ensued. [1]

Similarly, in 2021, Colonial Pipeline prioritized cybersecurity investments after a ransomware attack targeting its critical infrastructure. Recognizing its vulnerability to extortion attempts by financially motivated threats, it implemented stronger endpoint security measures and proactively engaged in incident response training. In this case, understanding the attacker's motivation allowed Colonial Pipeline to bolster its defenses and mitigate the impact of a subsequent attack, minimizing operational disruption and financial losses. These real-world examples demonstrate the crucial role of attacker motivation in building resilient security postures. By factoring in diverse motivations, organizations can anticipate a wider range of threats and proactively strengthen their defenses against even the most unexpected assaults. [2]

These two cases illustrate the critical difference between threat awareness tailored to your organization's specific landscape and a generic security approach. Understanding your unique threat profile fuels robust, contextually relevant security decisions, enabling you to proactively mitigate vulnerabilities exploited by attackers.

Next, we'll delve deeper into the diverse attacker arsenal and dissect how effectively analyzing its composition can inform strategic security choices.

A hacker's toolkit and its evolution

Modern attackers leverage diverse tools and techniques to exploit vulnerabilities and infiltrate systems. Imagine being a security architect protecting a fortress, but your enemy wields an ever-evolving repository. That's the reality of battling today's hackers, whose toolkit ranges from prefabricated exploit kits to bespoke zero-day creations. Let's crack open this digital armory and understand the weapons at play.

On the lower end, you'll find **exploit kits**: mass-produced bundles targeting common software vulnerabilities. Think of them as IKEA furniture for attackers – easy to assemble, even for beginners. These kits automate much of the hacking process, making it accessible to anyone with enough cash and malicious intent. While not the most sophisticated option, they pose a serious threat due to their sheer volume and ability to exploit unpatched systems. These automated toolkits serve as a potent resource, commonly employed by individuals with limited technical expertise (script kiddies). However, when coupled with sufficient resources, sophisticated attackers can effectively leverage these toolkits to identify and exploit the initial foothold in a system. [3]

The natural next step up would be **malware-as-a-service** (**MaaS**) platforms. Imagine subscribing to Netflix, but instead of movies, you get ready-to-deploy malware. These platforms offer a buffet of malicious tools – ransomware, keyloggers, botnet builders – with pay-as-you-go pricing and even customer support! This democratizes cybercrime, allowing even less skilled actors to launch sophisticated attacks. These are increasingly growing in popularity as the sheer number of options increases, giving motivated attackers the fuel to launch large-scale attacks.

Finally, we reach the pinnacle of hacker ingenuity: custom-crafted **zero-day** attacks. Think of these as bespoke suits, tailored to exploit specific vulnerabilities in your software or hardware. They are often the work of highly skilled individuals or well-funded groups, targeting high-value assets such as government networks or critical infrastructure. They demand significant resources and expertise, but their stealth and effectiveness make them a top concern for security professionals.

> **Zero-day exploits**
>
> A zero-day exploit refers to a cyberattack that takes advantage of a previously unknown or undisclosed vulnerability in software, hardware, or firmware. The term "zero-day" indicates that the developers have had zero days to address and patch the vulnerability, making it particularly potent for attackers. Since the targeted entity is unaware of the flaw, there are no patches or defenses in place to mitigate the exploit, giving attackers a significant advantage. Zero-day exploits are highly valuable and often used for targeted attacks or in the initial stages of widespread attacks before security patches can be developed and deployed. [4]

As defensive controls increasingly integrate into the early stages of the software development life cycle, the cost of zero-day exploits is on the rise. For instance, a zero-day exploit targeting a **remote code execution** (**RCE**) vulnerability in the Chrome browser is reported to command a price ranging from $500,000 to a whopping $1 million. While elevated security standards contribute to these escalating prices on the dark web, it's essential to recognize that such amounts, while significant, remain within the financial reach of state-sponsored attackers. [5]

Hackers, like any good entrepreneurs, understand the value of iteration and refinement. Their "innovation pipeline" isn't fueled by venture capital but by stolen data, compromised infrastructure, and vulnerabilities found in the vast pool of open source code. These spoils of past conquests become the building blocks for future attacks, evolving through a twisted cycle of refinement.

Stolen credentials from data breaches can be repackaged to access new targets, while compromised cloud resources become staging grounds for more potent malware campaigns. Open source libraries, riddled with hidden vulnerabilities, are readily adapted into attack frameworks, granting even novice hackers access to powerful offensive capabilities. Each successful exploit feeds back into the pipeline, fueling the development of even more effective tools and techniques. This perpetual cycle of weaponization and exploitation keeps security professionals on their toes, constantly anticipating the next twist in the attackers' innovation pipeline.

This pipeline isn't just about raw data; it's about knowledge and expertise. Hacker forums buzz with detailed post-mortem analyses of past attacks, sharing lessons learned and best practices for future exploits. This collaborative environment ensures that advancements made by one hacker quickly benefit the entire cybercriminal community, further accelerating the evolution of their digital weaponry. It is almost like a parallel universe to security professionals. [6]

So, how do we defend against this evolving arsenal? The answer lies in a layered approach. Patching vulnerabilities diligently, deploying advanced detection systems, and fostering a culture of cybersecurity awareness are all crucial. These concepts have been discussed in earlier chapters. Now, we can strategically assemble these various components, akin to constructing a resilient defensive fortress using a set of well-coordinated Lego pieces.

Understanding the attacker's business model

As cyber threats have evolved, so has the professionalization of hacking operations. No longer confined to isolated individuals or loosely organized groups, modern attackers often adopt sophisticated business models. These models involve strategic resource allocation, carefully planned operations, and a keen awareness of market dynamics. The rise of professional hacking organizations signals a shift toward more structured, profit-driven cybercriminal enterprises.

Imagine a shadow economy with its own currency, marketplaces, and even customer support. That's essentially what modern cybercrime has become. The initial attack might be the flashy headline, but the real story lies in the financial journey:

- **Data exfiltration**: Stolen information, from credit cards to medical records, is packaged and sold on underground marketplaces such as "dark web" forums. Imagine a digital bazaar where stolen goods change hands anonymously, often in exchange for cryptocurrency such as Bitcoin.

- **Cryptocurrency laundering**: These digital coins, while difficult to trace, aren't exactly untouchable. Hackers employ complex laundering schemes, often involving shell companies and offshore accounts, to convert their ill-gotten gains into "clean" money.

- **Ransomware cash-out**: Ransomware, where attackers encrypt your data and demand payment for its release, has become a lucrative business. Cybercriminals operate call centers and "customer support" channels dedicated to negotiating ransoms and facilitating payment through anonymous channels.

Moreover, professionalized hacking is becoming more and more common in the modern world. This isn't some ragtag group of teenagers in their basements anymore. Cybercrime has evolved into a professionalized industry, with attackers operating like well-oiled businesses. They do the following:

- **Specialize**: Gone are the days of the jack-of-all-trades hacker. Today, cybercriminals specialize in specific areas, from malware development to social engineering, forming highly skilled teams with defined roles and responsibilities.

- **Invest in resources**: Cybercrime is big business, and attackers reinvest their profits into cutting-edge tools, zero-day vulnerabilities, and even exploit development kits. They see it as an investment in their future success.

- **Embrace outsourcing**: Just like any corporation, hackers outsource tasks. Need custom malware written? There's a freelancer for that. Want to launch a DDoS attack? Hire a botnet-as-a-service provider. This collaborative approach maximizes efficiency and minimizes risk.

Demystifying the economics of cybercrime isn't just an academic exercise. It grants us invaluable insights into how attackers prioritize their targets and optimize their operations. This, in turn, empowers us to make smarter choices about our own security posture.

By understanding what motivates our digital adversaries, we can focus our defenses on the assets most likely to attract their attention. Is it our customer data, intellectual property, or critical infrastructure? Identifying the crown jewels in our digital kingdom allows us to concentrate our resources on fortifying those defenses, rather than scattering them across every potential entry point.

Furthermore, a nuanced understanding of attacker economics helps us distinguish between genuine threats and mere noise. Not all attacks are created equal. Knowing which cyber scams offer the highest return on investment for attackers allows us to prioritize responses, focusing on threats that pose the most imminent and significant risk, while filtering out the low-hanging fruit that hackers fling at everyone just in case.

Ultimately, viewing security through the lens of an attacker's business model allows us to shift from a reactive to a proactive stance. By anticipating their moves and understanding their motivations, we can allocate resources strategically, build more resilient defenses, and ultimately make it harder and less profitable for them to target our valuable assets. It's a game of chess, but by understanding both sides of the board, we can checkmate their malicious ambitions before they even make a move. It is important to recognize that we will continue to find new vulnerabilities and new ways of breaking into systems; the goal is not to build a 100% secure system (which is quite unachievable) but to make it reasonable.

Now, let's turn our attention to a dedicated group of threat actors.

Advanced persistent threats (APTs)

In the realm of cybersecurity, APTs represent a formidable adversary, distinguished by their sophisticated methodologies and persistent nature. APTs often originate from well-resourced threat actors, such as nation-state-sponsored groups or organized cybercrime entities. Understanding the intricate workings of APTs is essential for defenders seeking to fortify their systems against these persistent and stealthy adversaries. We will explore each phase of an attack to better understand the inner workings of APTs:

1. **Initial infiltration**: APTs excel at gaining a foothold in your network undetected. They leverage diverse techniques, from zero-day exploits in web applications to spear-phishing emails tricking employees into unwittingly installing malware. Notorious examples include Operation Aurora (2009) [7], where Chinese hackers exploited Adobe vulnerabilities to infiltrate Google and other tech giants, and NotPetya (2017) [8], where a seemingly mundane software update spread like wildfire, crippling government agencies and businesses across the globe.

2. **Establishing a foothold**: Once inside, APTs deploy persistence mechanisms to maintain their access. They might implant backdoors, hide malware within legitimate software, or hijack administrative accounts. The Sony Pictures hack (2014) showcased this tactic when North Korean attackers leveraged compromised credentials to wreak havoc, leaking sensitive data and even disrupting movie releases.

3. **Lateral movement and reconnaissance**: With a foothold established, APTs move stealthily through the network, mapping its layout and identifying high-value targets. Tools such as network sniffers and credential stealers help them harvest sensitive information, including usernames, passwords, and internal documents. The Stuxnet worm (2010) [9] is a chilling example that infiltrated Iran's critical infrastructure, exploiting vulnerabilities in Siemens industrial control systems to disrupt uranium enrichment facilities.

4. **Exfiltration and impact**: Finally, APTs capitalize on their ill-gotten gains by exfiltrating assets. Stolen data may be used for espionage, sold on the black market, or publicly leaked to achieve political or financial goals. Often, attackers blend exfiltration with destructive actions, as seen in the WannaCry ransomware outbreak (2017), where encrypted data became the bargaining chip for extortion payments.

Another notable APT group is APT29 [10], also known as Cozy Bear, associated with Russian state-sponsored cyber activities. This group gained global attention for its involvement in the breach of the **Democratic National Committee (DNC)** in 2016. APT29 utilized sophisticated spear-phishing campaigns and custom malware, demonstrating a high level of technical prowess and persistence.

Over the years, APTs have evolved, with groups such as APT28 (Fancy Bear) and APT34 (OilRig) showcasing diverse tactics. APT28 has been linked to cyber operations attributed to Russian intelligence, while APT34, believed to be associated with Iran, has engaged in cyber espionage targeting the Middle East. These examples underscore the need for defenders to stay abreast of evolving APT strategies and implement proactive measures to counter these persistent and highly skilled adversaries. APTs can feel very intimidating, and rightfully so, as they are often backed by large nation-states and state-of-the-art facilities. However, making fundamental changes such as embracing a zero-trust, human-less, reproducible production environment, mandatory security policies around code review, and multi-factor authentication for critical changes can seriously thwart even the most sophisticated attackers.

Thinking like an attacker – Identifying weaknesses

Imagine a master strategist meticulously studying their opponent, identifying vulnerabilities before they're exploited. This is the essence of effective defense: adopting the attacker's perspective to pinpoint your weaknesses and fortify your digital walls. This section equips you with the tools and techniques to do just that. Remember, this is not about mimicking an attacker but about looking at the systems you are protecting and asking yourself: what could go wrong?

We'll begin by crafting a detailed profile of our potential adversaries. Understanding their motivations, capabilities, and preferred attack vectors allows us to anticipate their moves and prioritize our defenses accordingly. Think of it as building a threat landscape map, where each attacker type occupies a distinct terrain based on their characteristics.

Next, we'll venture into the heart of your infrastructure, mapping your exposed assets, such as valuable data and critical systems. Imagine shining a bright light on every corner of your network, identifying potential attack entry points and vulnerable areas. By understanding what attackers target, we can prioritize defenses where they matter most.

Once the target is clear, we'll delve into the world of vulnerability management. This vital process involves identifying and patching security flaws in your systems and applications. Think of it as plugging every crack and sealing every leak before an attacker can exploit them.

But defense isn't just about patching holes; it's about prioritizing risks. We'll analyze the criticality of your assets and the severity of potential vulnerabilities, ensuring we address the most pressing threats first. Imagine ranking each potential attack scenario based on its likelihood and impact, ensuring your precious resources are directed toward the most dangerous vulnerabilities. We will reference some materials from the earlier chapter on risk-driven defense strategies in this section.

Lastly, integrating threat intelligence into our defensive arsenal provides real-time insights into emerging threats, enabling proactive defense measures. This ongoing process involves gathering and analyzing information about emerging threats, attacker tactics, and new vulnerabilities.

By embracing this proactive approach, you transform from a passive target into a strategic defender. This section equips you with the critical skills and knowledge to think like an attacker, identify your vulnerabilities, and construct an impregnable defense against even the most determined adversaries.

Profiling potential adversaries

In earlier chapters, we provided a concise overview of the spectrum of attackers, spanning from script kiddies to state-sponsored APTs. Each category of attacker operates with distinct motivations, influencing their tactics and objectives. The composition of the attacker landscape varies significantly based on the industry in which an organization operates. To delve deeper into this critical aspect, we will revisit this spectrum, this time with a strategic emphasis on constructing a robust framework. The objective is to facilitate the identification of the threat groups that pose the most substantial risk to your organization. Let's look at a few such adversaries.

- **Financially motivated hackers**: A financially motivated hacker seeks to exploit vulnerabilities in digital systems for monetary gain.

  - **Industry**: Broad spectrum, with a focus on sectors with readily monetizable data (e.g., finance, healthcare)

  - **Data assets**: PII, financial information, and credit card data

  - **Public profile**: Often operate under the radar, utilizing readily available tools and exploit kits

  - **Goals**: Quick financial gain through ransomware, data exfiltration, and online fraud

  - **Resource potential**: Varied, ranging from lone individuals to organized crime groups

- **State-sponsored actors**: State-sponsored attackers are individuals or groups backed by governments who engage in cyber operations to further political, economic, or military objectives.

  - **Industry**: Strategic sectors such as infrastructure, government agencies, and critical technology

  - **Data assets**: Intellectual property, classified information, and national security secrets

  - **Public profile**: Often state-backed, operating with advanced tools and techniques

  - **Goals**: Espionage, political disruption, infrastructure sabotage, and information gathering

  - **Resource potential**: Considerable, backed by significant government resources and expertise

- **Hacktivists**: Hacktivists are individuals or groups who use hacking techniques to promote political or social causes and express their beliefs or grievances.

  - **Industry**: Targets that often reflect their ideological motivations (e.g., activism, political dissent)

  - **Data assets**: Symbolic targets such as government websites and corporate databases

  - **Public profile**: Openly vocal about their motives, attracting media attention

  - **Goals**: Disruption, raising awareness, and promoting a specific cause

  - **Resource potential**: Generally moderate, relying on readily available tools and **distributed denial-of-service (DDoS)** attacks

- **Malicious insiders**: Malicious insiders are individuals within an organization who misuse their access and privileges to intentionally harm the organization's systems, data, or reputation.

  - **Industry**: Any organization with disgruntled employees or state actors getting into an organization to take advantage of privileged access

  - **Data assets**: Confidential information, internal systems, and customer data

  - **Public profile**: Often difficult to identify, blending in with authorized personnel

  - **Goals**: Personal gain, revenge, disruption, or providing access to external attackers

  - **Resource potential**: Varies based on their level of access and knowledge of internal systems

- **Script kiddies**: Script kiddies are individuals who lack advanced technical skills but use readily available scripts or tools to launch simple and often indiscriminate cyberattacks.

  - **Industry focus**: Opportunistic, targeting vulnerabilities without specific industry preferences

  - **Data assets**: Varied, often seeking to disrupt systems for personal amusement

  - **Public profile**: Ranges from low to high, depending on the desire for recognition

  - **Goals**: Building reputation within their community, gaining recognition for their skills, and causing disruption

  - **Resource potential**: Limited technical expertise, relying heavily on pre-written scripts and tools created by others

- Additional notes:

  - Often motivated by curiosity and a desire to prove their abilities

  - Can be unpredictable and opportunistic, exploiting known vulnerabilities without fully understanding the implications

  - May unintentionally cause significant damage due to their lack of experience and caution

  - Can serve as a gateway for more sophisticated actors, who may leverage their initial access to launch more targeted and damaging campaigns

By understanding these diverse attacker profiles, you can prioritize vulnerabilities based on their alignment with your industry, data assets, and public profile. Calculate the threat posed by each attacker profile. Combine their motivation, their capabilities, and your vulnerability posture to prioritize defensive investments. Remember, this is not an exhaustive list and emerging threats constantly reshape the landscape. Continuous threat intelligence gathering and monitoring remain crucial to maintaining a robust defense.

Mapping and hunting exposed assets

This goes back to the idea of protecting what you know. It is going to be almost impossible to protect an asset that an organization is not aware of. This section covers how we can utilize techniques such as attack trees, given the information about potential threat actors, and find weaknesses in our systems proactively.

The first step involves prioritizing our most valuable assets. Identify the data and systems crucial to your organization's operations, such as customer databases, financial records, and intellectual property. Consider which assets align with the motivations of your identified attacker profiles – financial hackers will naturally gravitate toward monetizable data, while state-sponsored actors might target confidential government information.

Next, let's trace the potential "attack highways" leading to these critical assets. Imagine you are tasked with attacking the system, analyzing entry points such as exposed public-facing applications, outdated software vulnerabilities, or even misconfigured network devices. Consider how compromised user accounts or insecure access controls could provide footholds within your network. Remember, adversaries often favor the path of least resistance, so focus on vulnerabilities with readily available exploits or those requiring minimal technical expertise.

> **Note**
>
> In certain scenarios, resource constraints may limit the financial and temporal investment available for defending a system. In such instances, the DiD strategy emerges as an invaluable ally. During the process of threat modeling, it becomes crucial to acknowledge that certain components of a system will inherently possess vulnerabilities. Incorporating this assumption into the foundational aspects of the system's defense strategy bolsters its overall resilience, thus making the task of an attacker significantly complicated.

By charting these potential attack paths, you gain invaluable insights into your security posture. You identify the weak links in your digital chain, allowing you to prioritize patching critical vulnerabilities, implement stricter access controls, and fortify perimeter defenses. This proactive approach transforms you from a passive fortress waiting to be breached into a vigilant guardian anticipating and thwarting potential intrusions before they cause damage.

Vulnerability management and patch prioritization

Now that we have explored the attack perspective to find weaknesses, the next step is to prioritize which vulnerabilities to act on first. Vulnerability management is a critical facet of maintaining a robust defense against cyber threats. It involves the systematic identification and prioritization of vulnerabilities within a system or network.

The cornerstone of this process is discovery. Leverage comprehensive vulnerability scanners and **security information and event management (SIEM)** systems to cast a wide net across your entire infrastructure. Scan applications, operating systems, firmware, and network devices, leaving no stone unturned in your quest for weaknesses. Remember, attackers exploit not just well-known vulnerabilities, but also zero-day flaws and less publicized exploits.

Once identified, vulnerabilities need to be categorized and prioritized. Assessing their severity requires considering multiple factors: exploitability, the potential impact on your specific data and systems, and the threat landscape targeting your organization. For example, a critical vulnerability in your customer database exploited by state-sponsored actors carries a far greater risk than a low-impact flaw in a non-essential application. Utilize vulnerability scoring systems, threat intelligence feeds, and internal risk assessments to prioritize patching, focusing your resources on the most immediate and impactful threats. We have extensively discussed risk-based prioritization of threats in *Chapter 1*, so please refer to them for a refresher.

Once you've prioritized, finally, comes the crucial act of patching. Apply security updates promptly, keeping your software and systems up-to-date with the latest fixes. Automate patching wherever possible to minimize the window of vulnerability. Utilize rollback plans and thorough testing procedures to mitigate the risk of patch-related disruptions. Remember, the time from vulnerability disclosure to exploit development is often short, so timely patching is a critical line of defense.

Threat intelligence for indicators of compromise (IoCs)

So far, we've put on the attacker's mask, mapping vulnerabilities and charting potential assault paths. Now, we flip the script, becoming proactive hunters wielding the tools of threat intelligence and **indicators of compromise (IoCs)** to identify breaches before they blossom into full-blown disasters.

Think of IoCs as digital footprints – breadcrumbs left behind by adversaries as they navigate your network. These can be suspicious file hashes, unusual network traffic patterns, and even specific malware signatures. By actively searching for these clues, we become forensic investigators meticulously combing the scene of a potential crime.

Advanced hunting techniques elevate this beyond mere clue-seeking. By leveraging threat intelligence feeds gleaned from industry sources, security researchers, government agencies, and even the dark web marketplace, we identify emerging threats, attacker TTPs, and the latest IoCs employed by adversaries. This empowers us to proactively hunt for these indicators within our own infrastructure, identifying even previously unseen threats lurking in the shadows.

In 2021, Microsoft's proactive threat hunting, fueled by intelligence on the SolarWinds supply chain attack, allowed it to detect intrusion within its own network before significant damage was inflicted. Similarly, analyzing suspicious network traffic patterns and IoCs associated with ransomware gangs allowed defenders to disrupt attacks and mitigate data exfiltration attempts before critical systems were compromised. In my personal experience, we were able to thwart the infamous WannaCry ransomware from impacting our organization's network by staying on top of threat feeds and making a simple tweak in our **web application firewall (WAF)** rule.

By thinking like an attacker during threat hunting, we gain invaluable insights into their preferred tools, techniques, and target selection. This enables us to anticipate their moves, identify weaknesses they might exploit, and proactively close those gaps in our defenses. Imagine defenders constructing sandcastles not after the tide rises, but before the first wave even approaches.

So far, we have explored the attacker perspective quite extensively. In the next section, we will shift toward specific TTPs that these attackers utilize.

Understanding TTPs

To truly solidify our defenses, we must not only understand who might attack us, but also how they might do it. This section unlocks the secrets of the attacker's arsenal, delving into the world of TTPs. Understanding these TTPs is crucial for defenders as it enables them to identify and counter potential threats effectively. Common patterns in TTPs often provide insights into the modus operandi of various threat actors, offering a foundation for analyzing and categorizing cyber threats.

Understanding common TTP patterns is the first step. We'll identify recurring elements across different attack types, such as the initial reconnaissance phase where attackers gather information about your systems, the exploitation stage where they leverage vulnerabilities to gain access, and the post-exploitation phase where they move laterally, steal data, and maintain persistence within your network. Each of these stages presents its own unique set of TTPs, from phishing emails and watering hole attacks in reconnaissance to privilege escalation techniques and zero-day exploits in exploitation.

But attackers aren't content with just gaining access; they strive to remain undetected and maintain a foothold within your network. We'll explore common persistent mechanisms, such as hidden backdoors, malware disguised as legitimate files, and clever techniques for evading detection by security tools. Understanding these methods allows us to implement appropriate countermeasures, such as **endpoint detection and response** (**EDR**) tools and advanced network monitoring, to identify and disrupt attacker persistence before they can inflict significant damage.

By demystifying the attacker's playbook, we gain a critical advantage. We can anticipate their moves, prioritize vulnerabilities based on the TTPs employed by relevant threats, and design defenses that effectively thwart their attempts at intrusion, persistence, and data exfiltration. This section equips you with the knowledge and tools to transform from a passive target into a proactive guardian, building defenses that anticipate and outmaneuver even the most cunning adversaries. Let's begin with understanding TTPs.

Understanding TTPs and common patterns

We have thoroughly examined various attacker profiles and the motivations driving their attacks. Now, let's delve into portraying the common TTPs employed by these attackers across the attack life cycle stages, revealing their preferred tactics and techniques at each phase to further enhance our understanding. By understanding how specific adversaries operate, we can anticipate their moves, prioritize vulnerabilities more effectively, and strengthen our defenses against targeted attacks.

- **Financially motivated hackers**:

  - **Reconnaissance**: Phishing campaigns, social engineering, web scraping, vulnerability scanning

  - **Exploitation**: Ransomware deployment, zero-day exploits, brute-force attacks, SQL injection

  - **Post-exploitation**: Data exfiltration (PII, financial information), credential theft, cryptocurrency mining, botnet integration

- **State-sponsored actors**:

  - **Reconnaissance**: Advanced spear phishing, zero-day exploitation, supply chain attacks, social media monitoring

  - **Exploitation**: Custom malware, fileless techniques, privilege escalation, zero-day vulnerabilities

  - **Post-exploitation**: Espionage (sensitive data exfiltration), infrastructure sabotage, disruption campaigns, information manipulation

- **Hacktivists**:

  - **Reconnaissance**: Website crawling, DDoS target identification, vulnerability scanning, social media monitoring

  - **Exploitation**: Defacement attacks, website vulnerabilities, DDoS attacks, social engineering

  - **Post-exploitation**: Website defacement, data deletion, disruption of critical services, spreading propaganda or messages

- **Malicious insiders**:

  - **Reconnaissance**: Internal reconnaissance tools, privileged access exploitation, social engineering of colleagues

  - **Exploitation**: Internal phishing, data manipulation, system configuration changes, exploiting existing access

  - **Post-exploitation**: Data exfiltration (trade secrets, customer data), financial fraud, sabotage of internal systems, selling access to external attackers

Remember, these are just generalized examples, and individual attackers within each type might exhibit variations in their TTPs. However, by understanding the common techniques employed by each category, we can gain valuable insights into their motivations, target selection, and preferred attack paths. This knowledge empowers us to prioritize specific vulnerabilities based on the threat landscape relevant to our organization, allocate resources effectively for targeted defense, and implement security controls that disrupt their attacks at various stages.

With this high-level understanding of common patterns, we will next dive a little deeper into different phases.

Exploitation techniques and vulnerability exploits

If we analyze the landscape of critical exploits over the last few decades, they exhibit a growing maturity that can be systematically categorized into distinct vulnerability classes. In this section, our goal is to construct a conceptual framework highlighting prevalent exploitation techniques employed by attackers. This approach is analogous to the deny-list approach, which has its inherent limitations. Nevertheless, this knowledge serves as a valuable instrument for recognizing emerging and intricate issues specific to your systems. To illustrate, consider the notorious Stuxnet worm, which exploited multiple zero-day vulnerabilities, including a Windows kernel exploit (CVE-2010-2743), showcasing the potency of such exploitation techniques. This retrospective examination equips defenders with insights to enhance their security posture against evolving threats. We will explore the tools and tactics that adversaries employ to breach our defenses, leveraging software vulnerabilities, escalating privileges, and navigating our systems with malicious intent.

Memory corruption exploits

These techniques target flaws in software that allow attackers to inject malicious code into memory. Once executed, this code can bypass security controls, steal data, or take control of the system. Common examples include the following:

- **Buffer overflows**: Writing more data to a buffer than it can hold, overwriting adjacent memory, and potentially injecting malicious code (e.g., Heartbleed vulnerability) [11]

- **Heap overflows**: Like buffer overflows, but targeting dynamic memory allocation on the heap (e.g., DirtyPipe vulnerability)

- **Code injection**: Directly injecting malicious code into memory through vulnerabilities in applications or libraries (e.g., SQL injection, cross-site scripting)

RCE exploits

These attacks allow attackers to execute arbitrary code on a target system remotely and gain complete control over its resources. Common examples include the following:

- **Zero-day exploits**: Leveraging undiscovered vulnerabilities in software before a patch is available (e.g., EternalBlue for Windows SMB) [12]

- **Web application vulnerabilities**: Exploiting flaws in web servers, frameworks, or plugins to execute code on the server (e.g., Struts2 framework vulnerabilities)

- **API vulnerabilities**: Compromising application interfaces to gain access to underlying systems and data (e.g., Log4Shell vulnerability) [13]

Privilege escalation techniques

Once attackers gain initial access, they often seek to elevate their privileges to gain administrator-level control over the system. Common methods include the following:

- **Lateral movement**: Exploiting vulnerabilities or misconfigurations to move from compromised systems to more privileged ones (e.g., pass-the-hash attacks)

- **Exploiting local vulnerabilities**: Leveraging flaws in local applications or system services to escalate privileges within the compromised system (e.g., Windows local administrator password reset vulnerability)

- **Social engineering**: Tricking users into granting administrative privileges or revealing sensitive information (e.g., phishing attacks)

Remember, this is not an exhaustive list, and new vulnerabilities and exploit classes emerge constantly. Staying informed about the latest threats and patching critical vulnerabilities promptly is crucial in mitigating these attacks.

Persistence mechanisms

Attackers employ an array of techniques to prolong their presence within a network and subvert systems without raising alarms. Once inside, attackers rarely want to leave. This section delves into the cunning world of persistence mechanisms and the techniques attackers employ to maintain a foothold within your network, remain undetected, and evade security measures.

One common tactic involves establishing **backdoors** – clandestine entry points that bypass traditional authentication and security controls. These can be hidden within compromised applications, system files, or even network configurations. The infamous SolarWinds supply chain attack, for instance, utilized backdoored software updates to provide attackers with long-term access to victim networks. Another notable example includes the Shadow Brokers' use of the DoublePulsar backdoor in the infamous Equation Group toolkit, emphasizing the real-world impact of such techniques. [14]

Another stealthy trick is deploying **rootkits**, software designed to mask malicious activity and modify system operations. They hide processes, files, and network connections, effectively operating in the shadows beneath the radar of security tools. The Stuxnet malware, targeting Iranian nuclear facilities, employed a sophisticated rootkit to remain invisible while sabotaging critical infrastructure. Rootkits are often very hard to deploy with modern defense controls but are also harder to detect. These stealthy tools often replace or manipulate essential system files, allowing attackers to execute malicious commands without detection.

Beyond physical footprints, attackers also utilize covert channels to steal data and communicate undetected. These channels can be embedded within seemingly innocuous data streams, such as DNS requests, ICMPs, or image files, making them difficult to discern from legitimate traffic. The Duqu malware, a relative of the infamous Stuxnet, utilized covert channels for **command-and-**

control (C2) purposes, underscoring the significance of comprehending these techniques in modern cybersecurity. In 2009, cybercriminals exfiltrated data from RSA by hiding it within images used in employee training materials.

Understanding these persistence mechanisms is crucial to effective defense. We must implement security tools that scan for anomalous activity, monitor system changes, and detect hidden processes and network communications. Implementing strict application allow-listing, data loss prevention controls, and EDR solutions can further thwart their attempts to remain embedded within our networks.

One thing to think about is that attackers are constantly trying to get into our systems. It is very reasonable to assume that if your system has any path from the internet, someone is trying to get into it and it will be attacked. But if you take a step back, it becomes evident that attackers often employ similar techniques as they navigate from one system to another. This pattern underscores the strategic intent of attackers to conceal their methods, even after compromising a system. A defender's job is not only to try to prevent attacks from happening but also to build capabilities into systems to forensically analyze what went wrong.

In the next section, we will look deeper into common methodologies used by attackers to not only stay undetected while they are there, but to also remove any traces that they were ever there!

Evasion techniques and anti-forensics

In the perpetual cat-and-mouse game of cybersecurity, attackers employ a variety of evasion techniques to outsmart defenders and avoid detection. This section delves into the realm of evasion techniques that attackers employ to obscure their activities, manipulate logs, and vanish like digital ghosts. Understanding these tricks of the trade equips us to counter their maneuvers and shine a light on their hidden tricks.

A common tactic involves file and process obfuscation. Attackers modify malware signatures, encrypt payloads, and inject code dynamically to bypass traditional antivirus and intrusion detection systems. For instance, WannaCry ransomware employed multiple obfuscation techniques to evade initial detection, delaying its identification and mitigation efforts. Malware obfuscation is a very interesting topic; there are different methods including compression, encryption, and so on. [15]

Beyond masking their presence, attackers also manipulate the environment by employing anti-forensics tactics to erase their footprints and hinder investigations. This can involve altering timestamps, wiping logs, and disabling logging altogether. The NotPetya wiper malware, designed to disrupt critical infrastructure, notoriously overwrote system files and cleared event logs, making forensic analysis and recovery immensely challenging.

Recognizing these evasion techniques is critical to effective defense. Employing advanced security tools with multi-layered detection capabilities and leveraging behavioral analysis and anomaly detection can help unmask hidden threats. Maintaining comprehensive and tamper-proof logs, implementing data integrity solutions, and employing forensic readiness strategies further strengthen our defenses against stealthy adversaries.

Now let's look at some emerging trends that attackers are leveraging to stay undetected.

Living off the land attacks

In the world of **living off the land** (LOTL) attacks, adversaries hijack legitimate tools and processes built into your own systems to achieve their malicious goals. LOTL attacks represent a strategic shift in the tactics employed by attackers, as they capitalize on existing, legitimate tools and processes within a target environment. Unlike traditional attacks that may introduce custom malware, LOTL attacks leverage the built-in functionalities of operating systems and trusted applications, making them harder to detect. Attackers essentially operate within the boundaries of the target's infrastructure, minimizing the need for conspicuous activities that might trigger alarms.

The benefits to attackers are numerous. By leveraging already present tools in your environment, attackers bypass traditional security checkpoints and blend seamlessly with routine system activity. This makes detection exceptionally challenging, as malicious actions appear identical to normal system functions. Moreover, these attacks don't require additional malware downloads or complex exploits, further reducing their footprint and increasing their chances of success.

In real-world scenarios, LOTL attacks often involve the exploitation of PowerShell or other scripting languages that are native to the system. By executing malicious code through these legitimate scripting tools, attackers can evade traditional antivirus solutions and appear as ordinary system processes. This technique was notably employed in the PowerShell-based "PowerSploit" framework, which facilitates post-exploitation activities while leaving minimal traces, complicating forensic analysis.

The infamous Mimikatz tool, originally designed to extract passwords from memory, has been abused by attackers in countless LOTL campaigns to steal credentials and gain unauthorized access.

Defending against LOTL attacks requires a shift in perspective. Traditional signature-based detection falls short, as it struggles to distinguish the malicious use of legitimate tools from normal activity. Instead, defenders must focus on behavioral analysis, anomalous activity detection, and close monitoring of system usage. Employing tools that provide visibility into process execution, network traffic analysis, and endpoint behavior can help identify suspicious activity even when malicious code is absent.

Defensive countermeasures – Turning the tables

Throughout this chapter, we've worn the attacker's hat, delving into their TTPs, exploring their arsenals, and witnessing their innovative tactics. We've navigated the intricate stages of their campaigns, from initial reconnaissance to persistent footholds and covert maneuvers. But the journey doesn't end here. Armed with this newfound knowledge, it's time to shift gears, transform from passive targets into proactive defenders, and reclaim control of our digital landscape.

A proactive and strategic defense requires a mindset shift from reactive measures to a more anticipatory and adaptive approach. This final section empowers you to embrace an adaptive defense mindset. We'll move beyond static and signature-based detection, venturing into the realm of proactive threat hunting, intelligence-driven countermeasures, and continuous security adaptation. By understanding attacker TTPs, leveraging threat intelligence feeds, and implementing advanced security tools, we can turn the tables on our adversaries, transforming from reactive victims into strategic defenders, actively thwarting their attempts, and safeguarding our precious digital assets.

Mindset shift in defense

For too long, defenders have operated in a reactive mode, scrambling to patch vulnerabilities and mitigate threats after they materialize. This section advocates for a fundamental mindset shift: embracing an adversarial mindset to anticipate attacker moves and proactively fortify our defenses. Imagine yourself walking in the shoes of an attacker, analyzing your systems for exploitable weaknesses, identifying critical assets, and charting potential attack paths. Make **offensive security** a core part of your defense strategy.

By adopting this perspective, we gain invaluable insights into the attacker's world. We identify vulnerabilities they might exploit, prioritize patching based on their preferred TTPs, and anticipate their attempts at gaining access, establishing persistence, and exfiltrating data. This proactive approach allows us to transform from passive guardians waiting for the next breach to vigilant hunters actively seeking and disrupting threats before they can inflict damage.

The benefits of this adversarial mindset are numerous. It fosters a culture of continuous security improvement, driving us to constantly evaluate our defenses, identify weaknesses, and implement mitigating controls. It empowers us to make informed decisions about resource allocation, focusing our efforts on the vulnerabilities most likely to be targeted by relevant threats. Ultimately, it allows us to turn the tables on attackers, shifting the balance of power in our favor and creating a more resilient security posture. Remember, it all starts with building security into your system, rather than treating it like a checklist.

Building adaptive defenses

The foundation of adaptation lies in continuous monitoring and threat intelligence integration. Leverage SIEM systems and EDR solutions to gain real-time visibility into system activity, identify anomalous behavior, and detect potential threats. Integrate threat intelligence feeds from industry sources, government agencies, and security researchers to stay informed about emerging attack patterns and vulnerabilities. This continuous flow of information allows you to proactively adjust your defenses, prioritize vulnerabilities based on real-world threats, and anticipate attacker tactics before they impact your organization.

Beyond passive monitoring, implement dynamic security controls that can automatically respond to evolving threats. Utilize security automation tools to streamline incident response, patch vulnerabilities promptly, and isolate compromised systems. Consider employing deception technologies to lure attackers into controlled environments (honeypots), sandbox suspicious files for analysis, and leverage machine learning algorithms to detect novel threats and anomalous activity patterns. Remember, agility is key in the cyber battlefield. By building defenses that can learn, adapt, and respond automatically, you gain a significant advantage over adversaries who rely on static attack methods.

Strategic countermeasures

After peering into the attacker's toolbox, we turn the tables and use this hard-earned knowledge to craft strategic countermeasures that thwart their attempts and safeguard our digital assets. We utilize the attacker's own playbook against them, anticipating their moves and disrupting their operations before they can gain a foothold.

By understanding common attack tactics, you can strategically prioritize your defenses. Focus on patching vulnerabilities aligned with the TTPs employed by threats relevant to your industry or organization. For instance, if financially motivated attackers are a primary concern, prioritize patching vulnerabilities exploited in ransomware campaigns. Implement security controls that specifically address these attack methods, such as multi-factor authentication to impede credential theft or email security solutions to counter phishing attempts.

Remember, threat intelligence is your friend. Leverage threat feeds and advisories to stay informed about emerging attack patterns and vulnerabilities. Proactively address these threats before they appear on your doorstep, patching critical vulnerabilities and adjusting your security posture accordingly. By strategically aligning your defenses with the attacker's common patterns, you stay one step ahead.

Summary

With a robust DiD framework as our foundation, our focus transitioned to comprehending the unique perspective of the potential threats our systems face. This insight serves as a potent tool in shaping our defense strategy. It is crucial to clarify that our objective wasn't to emulate real-world attackers but rather to grasp the distinct motivations driving their actions. This understanding allows us to scrutinize our systems from a fresh viewpoint. We delved into the array of tools employed by attackers to breach our defenses, identifying recurring patterns through historical analysis that provide valuable insights into potential vulnerabilities within our systems.

Our exploration extended to various attacker profiles and their common targets, equipping us with the knowledge to shift gears from being passive targets to being proactive defenders. Embracing an adversarial mindset became pivotal, enabling us to anticipate potential attacker maneuvers and fortify our defenses accordingly. The concept of adaptive defenses took center stage, leading to the development of security controls that continually evolve and adapt to emerging threats. This evolution is fueled by continuous monitoring, the integration of threat intelligence, and the implementation of dynamic security measures.

Key takeaways

- Thinking like an attacker does not mean imitating one. An adversarial mindset is an essential skill to foster in your workforce.

- Peering into an attacker's playbook reveals vulnerabilities and strengthens your defenses.

- By learning common TTPs, you can anticipate attack moves, evolve your defenses, and become a vigilant hunter.

- Leverage monitoring, threat intel, and dynamic controls to adapt to the ever-shifting threat landscape.

- Align your defenses with attacker TTPs, prioritize vulnerabilities, and outsmart their strategies.

- Knowledge is your weapon, and by gaining insights into attacker behavior, you can build robust defensive strategies.

Utilizing the distinctive insights gained from the attacker's perspective, we will progress to the next chapter with the goal of crafting a framework to apply this knowledge in safeguarding organizations against individualized modern threats.

Congratulations on getting this far! Pat yourself on the back. Good job!

Further reading

To learn more about the topics that were covered in this chapter, take a look at the following resources:

- [1] Equifax data breach: `https://www.ftc.gov/enforcement/refunds/equifax-data-breach-settlement`

- [2] Attack on Colonial Pipeline: `https://www.cisa.gov/news-events/news/attack-colonial-pipeline-what-weve-learned-what-weve-done-over-past-two-years`

- [3] AutoRecon GitHub: `https://github.com/Tib3rius/AutoRecon`

- [4] Zero-day: `https://en.wikipedia.org/wiki/Zero-day_(computing)`

- [5] Private zero-day: `https://www.bitdefender.com/blog/hotforsecurity/1-million-private-zero-day-bounty-reward-for-ios-9/`

- [6] Darknet Diaries: `https://darknetdiaries.com/episode/132/`

- [7] Operation Aurora on Google: `https://www.youtube.com/watch?v=przDcQe6n5o`

- [8] Untold story of NotPetya: `https://www.wired.com/story/notpetya-cyberattack-ukraine-russia-code-crashed-the-world/`

- [9] Stuxnet explained: `https://www.csoonline.com/article/562691/stuxnet-explained-the-first-known-cyberweapon.html`

- [10] APTs: `https://attack.mitre.org/groups/G0016/`

- [11] Buffer overflow playground: `https://tryhackme.com/room/bufferoverflowprep`

- [12] MS17-010: `https://www.exploit-db.com/exploits/42315`

- [13] Log4Shell vulnerability: `https://www.ncsc.gov.uk/information/log4j-vulnerability-what-everyone-needs-to-know`

- [14] ShadowBroker DoublePulsur: `https://www.rapid7.com/security-response/doublepulsar/`

- [15] Malware obfuscation: `https://ieeexplore.ieee.org/abstract/document/5633410`

5
Uncovering Weak Points through an Adversarial Lens

Building on the valuable knowledge gained from exploring attacker methodologies in the previous chapter, we will now shift gears toward actionable insights. This chapter empowers you to transform awareness into concrete steps, identifying and addressing your organization's vulnerabilities.

Think of it as a methodical examination, illuminating potential weaknesses within your security posture. Through organizational risk profiling, we'll equip you with tools to assess your unique threat landscape. This involves considering industry-specific risks, identifying your most valuable assets, and mapping out potential attack vectors. This self-evaluation becomes your roadmap, helping you prioritize resources and focus your efforts on the areas most critical to your organization's security health.

After that, we will delve into the dynamic world of **Defense in Depth (DiD)** with Red/Blue Teams. We will learn how a red team of ethical hackers plays the role of attackers, launching targeted assaults on your defenses. These exercises, mimicking real-world attacks, expose vulnerabilities from an adversary's perspective. The invaluable insights gained allow you to patch weaknesses before real attackers can exploit them.

But our journey doesn't stop there. We'll explore how to cultivate an adversarial mindset within your company. By fostering a culture where continuous security improvement and proactive threat hunting become embedded in your organization's DNA. Empowering your team to think like attackers becomes a powerful weapon in your cyber defense arsenal.

The objective here is to develop a systematic methodology for identifying attacker profiles with the highest threat potential to your organization or product. It aims to furnish a framework for action, constructing a DiD strategy tailored to individual organizational risks. This strategy incorporates effective controls designed to enhance the overall security posture and deliver tangible value.

In this chapter, we're going to cover the following main topics:

- Profiling organizational risks
- DiD for security organizations with red/blue teams
- Targeted approach to controls and strategies

Let's get started!

Profiling organizational risks

In the previous chapter, we put on an attacker mask, peering into their world and understanding their motivations, tactics, and tools. Now, you will leverage this newfound knowledge to delve into your organization's unique security landscape. This section empowers you to embark on a risk profiling journey, but with a crucial twist: we'll approach it through the adversarial lens.

The nature of data that an organization handles influences the primary threat actors targeting its systems. For instance, widely used software or services such as Microsoft Office and Google Chrome may be subject to ongoing threats from state-sponsored attackers [1]. To delve deeper into what this really means, when defending a system, it is imperative for the defender to pose a fundamental question: what are the potential consequences if an adversary successfully compromises our software or systems? This question serves as a precursor to the discussions in the previous chapter, *Chapter 4, Understanding the Attacker Mindset*. By understanding the implications of unauthorized access to our systems, we uncover the key to comprehending the distinctive threat landscape that an organization confronts.

Imagine yourself not just assessing risks but actively searching for vulnerabilities with the same question: what is at stake? Instead of relying solely on traditional risk assessment frameworks, you'll harness the attacker's perspective, identifying and prioritizing weaknesses they're most likely to exploit. This shift in mindset transforms passive risk assessment into a proactive exercise, empowering you to build a security roadmap fueled by DiD principles. By viewing your organization through the attacker's eyes, you gain an invaluable advantage: the ability to anticipate threats, prioritize defenses, and proactively fortify your security posture before adversaries can exploit your weaknesses.

This section equips you with the tools and techniques to transform risk profiling from a routine exercise into a strategic weapon. You'll learn how to leverage industry-specific threat intelligence, map potential attack vectors based on attacker TTPs, and prioritize vulnerabilities based on their criticality and potential impact. By the end of this section, you'll possess a clear roadmap, not just identifying risks but actively mitigating them through a robust and targeted DiD strategy informed by the attacker's playbook.

Organizational data profiling

The fundamental advice in security often revolves around the concept of minimizing data exposure. The idea is straightforward: without data, there is less to protect. However, in the realm of applications, which heavily rely on data for functionality, the challenge becomes balancing usability with security. Modern applications, driven by advancements in machine learning and artificial intelligence, are becoming increasingly data hungry [2]. The critical question persists: how can we optimize data usage within our software? One avenue is exploring strategies to reduce data dimensionality. For instance, when tracking user activity, considerations such as anonymizing collected data emerge as potential measures to mitigate risks.

As data holds a central position in an organization's risk landscape, the precise classification of this data becomes a critical priority. Now, we will outline a systematic, step-by-step approach to initiate organizational data profiling. Ideally, this process should be integrated into the workflow from the early stages of system development. However, it is important to note that these exercises can be undertaken at any point in the software life cycle. It becomes progressively challenging, though, as the age of the software increases. For instance, initiating such a process for a 15-year-old software system would be significantly more complex compared to a newly built system.

Building a data inventory

The first step is a comprehensive data inventory. Identify all data types your organization collects, stores, and processes. This includes customer information, financial records, intellectual property, and any other confidential data. Understand where this data resides (on-premises, cloud, or hybrid environments) and how it flows within your organization. Referring to the concept of data minimization, pose the question: what are the consequences if we were to delete this data? If the response indicates minimal impact, it may be prudent to consider the deletion of such data.

Categorizing your data

Now, begin by categorizing the data your organization deals with. Classify it based on its sensitivity, assigning different levels of importance to various types of information. For instance, personal customer details, financial records, or proprietary intellectual property might be deemed highly sensitive, while general operational data may be considered less critical. Not so fast, critical data, even if not inherently sensitive, could cause significant harm if compromised, such as financial records or operational data. Be mindful before discounting the criticality of operational information.

Identifying potential adversaries

Once your data is categorized, the next step is to consider who might be interested in this information. State-sponsored attackers might target organizations with valuable intellectual property, while financially motivated cybercriminals could focus on personal or financial data. Understanding the potential adversaries based on the nature of your data is crucial for developing targeted defense measures.

Consider revisiting *Chapter 4* to refresh your memories on the motivations and capabilities of potential adversaries, such as state-sponsored actors, cybercriminals, or hacktivists.

Tailoring your defenses

Finally, use your data insights to inform your security strategy. Prioritize security investments and resource allocation based on the data most at risk. Implement targeted security controls aligned with the specific threats you face. For example, if PII is a major concern, focus on strengthening data encryption and access controls. Remember, a one-size-fits-all approach won't suffice. By understanding your data and potential adversaries, you can tailor your defenses to effectively protect your most valuable assets. We will revisit this topic in greater detail toward the end of this chapter.

By following these steps, you transform data profiling from a static exercise into a dynamic threat mitigation strategy. Remember, your data map is a living document, requiring regular updates as your data landscape evolves. This ongoing vigilance ensures your defenses remain aligned with the ever-shifting threat landscape, safeguarding your organization's critical assets and data treasures.

Adversarial simulation

In this section, we will delve into an innovative technique to enhance organizational risk profiling. While adversarial simulation shares some concepts with the offensive security practices discussed in *Chapter 4*, this simulation is distinctive. It involves assuming that an attacker has already infiltrated certain parts of your system. Ask yourself a couple of critical questions: what controls does the system offer to thwart further penetration, and what is the organizational risk now that the attacker has gained access? This exercise proves invaluable for a security team aiming to construct a robust layered security approach. While there may be parallels with the concept of Netflix's **Chaos Monkey** [3] here, Chaos Monkey primarily focuses on system resiliency.

Traditional risk assessments often struggle to capture the true scope and impact of potential attacks. With adversarial simulation, we can experience the consequences of a breach firsthand, and design defenses that directly address the goals of likely adversaries.

Walking a mile in their shoes

Instead of passively assessing risks, adversarial simulation puts you in the attacker's shoes. Imagine conducting a virtual heist, using real-world attack techniques and tools to exploit vulnerabilities within your systems. This hands-on experience allows you to identify critical weaknesses, understand attacker motivations, and assess the potential impact of a successful breach.

Motivations matter

Understanding why attackers target your organization is crucial to designing effective defenses. Consider the motivations of potential adversaries – are they after financial gain, intellectual property theft, disruption, or data for further attacks? Analyzing real-world attacks and attacker profiles in your industry sheds light on their likely objectives.

Simulating the damage

During the simulation, map the attacker's path through your systems, mimicking their potential movements and objectives. Assess the damage they could inflict – data exfiltration, system disruption, reputational harm, or financial losses. Quantifying the potential impact helps you prioritize vulnerabilities and allocate resources effectively.

Building a risk profile with foresight

The insights gained from your simulated breach are invaluable. You'll identify critical weaknesses attackers are most likely to exploit, allowing you to prioritize organizational risk with much better clarity. By understanding attacker motivations, you can tailor your defenses to directly address their goals, making it significantly harder for them to achieve their objectives.

Note that the preceding outlined procedure operates at an abstracted level, foregoing the granular details of infiltrating systems through common TTPs, MITRE ATT&CK, or cyber kill chains [4]. The primary objective at this stage is to construct a conceptual map of doomsday scenarios, facilitating the precise categorization of risks that hold the utmost relevance.

Prioritizing risks with an attacker's mindset

Traditional risk assessments often rely solely on asset value and likelihood of occurrence to prioritize vulnerabilities. While valuable, this approach neglects a crucial dimension: the attacker's perspective. Fun fact: an attacker does not care how many compliance checkboxes an organization has checked. This section guides you to move beyond the numbers and integrate attacker motivations and capabilities into your risk assessment, leading to a more robust and targeted defense strategy.

Breaking free from the formula

Imagine an attacker prioritizing targets solely based on their monetary value, ignoring factors such as security measures or ease of access. Turn the tables, traditional risk assessments often make similar oversimplifications. By incorporating the attacker's perspective, you gain a more nuanced understanding of which vulnerabilities are most likely to be exploited.

Understanding "attacker ROI"

Consider the concept of attacker **return on investment** (**ROI**). Attackers prioritize targets based on the potential gain (financial, reputational, etc.) compared to the difficulty of exploitation. Analyze attacker motivations in your industry – are they after quick financial gains, intellectual property, or long-term data harvesting? An attacker will almost always choose the easiest path in!

Ease of exploitation matters

Don't just assess the potential impact of a breach; consider how easy it is for attackers to exploit a vulnerability. Outdated software, weak passwords, and misconfigured systems offer attackers easier entry points compared to well-patched and secured systems.

Prioritizing with precision

By integrating attacker motivations, capabilities, and potential ROI into your risk assessment, you can prioritize vulnerabilities with greater precision. Focus on patching critical weaknesses that offer attackers high potential gain with low exploitation difficulty. This targeted approach maximizes your security investments and significantly strengthens your overall defense posture.

To conclude, let's encapsulate the key takeaways from this section through a comprehensive mind map, showcasing deliverables from each step.

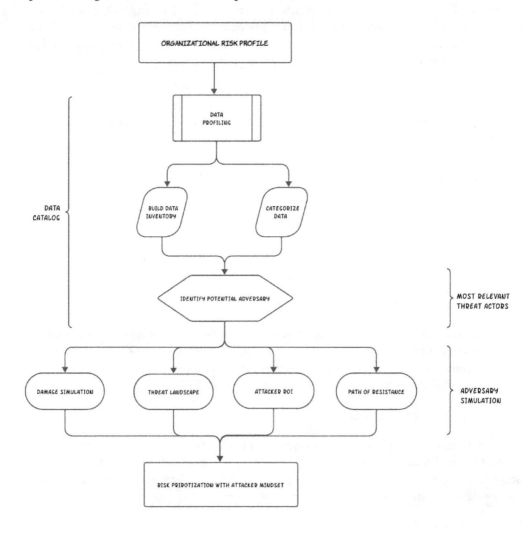

Figure 5.1 – Organization risk profile framework

This section introduced the concept of multi-dimensional risk assessment, integrating insights from the adversarial landscape into the risk management model. The evolution from traditional risk assessment methodologies to this attacker-centric mindset involves a strategic alignment with the adversary's perspective. This holistic approach acknowledges that the impact of a vulnerability is intricately tied to how enticing it appears to those seeking to exploit it. Consequently, the prioritization mechanism adapts to consider not only the technical severity of a vulnerability but also its appeal to potential attackers. This nuanced understanding allows organizations to focus resources on fortifying vulnerabilities that hold the greatest allure for adversaries, thereby enhancing the overall resilience of their security posture.

DiD for security organizations with red/blue teams

In the realm of cybersecurity, the perpetual battle between attackers and defenders necessitates a proactive and evolving approach. Strategies that rely on building defensive controls without methodically verifying their effectiveness often fall short in the face of the ever-changing threat landscape. This section introduces the **red team** and **blue team**, a powerful partnership that fuels continuous security improvement by replicating real-world attacks and strengthening your defenses from an adversarial perspective. This symbiotic relationship fosters a continuous cycle of improvement, ensuring that security controls remain robust and adaptive in the face of emerging threats.

Imagine a scenario where half of your security team (blue team) acts as skilled defenders, constantly striving to improve their detection and response capabilities. Meanwhile, the other half (a dedicated red team) plays the role of relentless attackers, actively seeking vulnerabilities and exploiting weaknesses within your systems. This continuous "war game" exposes real-world vulnerabilities, allowing your blue team to patch weaknesses, refine detection methods, and enhance their overall security posture. Read how Google benefits from having red teams to ensure the safety of their AI systems [5].

The domain of security is extensive, encompassing numerous moving components. Clearly delineating responsibilities not only enhances vulnerability detection but also contributes to the overall maturity of the security program. By adopting this adversarial lens, you move beyond simply measuring the effectiveness of existing security controls. Red team provides invaluable insights into how attackers might circumvent those controls, helping you identify critical blind spots and prioritize mitigation strategies. This proactive approach ensures your defenses stay ahead of the curve, ready to face even the most sophisticated adversaries.

In this section, we'll delve into the formation and operation of effective red and blue teams, highlighting their individual roles and responsibilities. We'll explore methodologies for conducting realistic red team engagements, ensuring valuable insights are gained without compromising real-world operations. Finally, we'll discuss how to analyze the results of these simulated attacks, translating them into actionable steps for continuous security improvement.

Building effective red/blue teams

Establishing red and blue teams within a security organization demands meticulous planning and thoughtful composition. The formation of these teams involves the strategic selection of skilled individuals possessing diverse expertise. The red team, comprising ethical hackers and penetration testers, brings an adversarial mindset, aiming to emulate real-world attackers. Their goal is to assess the effectiveness of existing security measures by attempting to breach systems using various attack vectors. On the other side, the blue team, comprised of defenders and security analysts, focuses on proactive monitoring, incident response, and fortifying defenses. The synergy between the red and blue teams creates a dynamic environment where offensive and defensive strategies converge.

The formation process includes defining clear roles, establishing communication channels, and fostering collaboration between the red and blue teams. This collaborative approach ensures that both teams understand each other's objectives and work toward a shared goal: strengthening the organization's security posture. The continuous interplay of red and blue teams not only enhances the organization's resilience but also fosters a culture of ongoing improvement, where lessons learned from each engagement contribute to refining security strategies and fortifying against evolving threats. Let's explore them one by one.

Red team

The red team represents the offensive arm of your security posture, playing the role of relentless adversaries seeking to exploit vulnerabilities within your systems. Their goal is not to cause harm but to expose weaknesses and push your defenses to their limits, ultimately leading to a more robust security ecosystem.

The ideal red team composition blends diverse skill sets. Look for individuals with expertise in the following:

- **Penetration testing**: Possessing the ability to identify and exploit vulnerabilities in networks, systems, and applications.

- **Social engineering**: Skilled in manipulating human behavior to gain access to information or systems.

- **Reverse engineering**: Able to analyze and understand the inner workings of software and exploit hidden vulnerabilities.

- **Security misconfigurations**: Adept at identifying and exploiting misconfigurations in security settings and protocols.

Red team activities must adhere to strict ethical guidelines. These include the following:

- **Clear authorization and scope**: All red team engagements must be pre-approved and have a defined scope to avoid unintended consequences.

- **Focus on discovery, not destruction**: The goal is to identify vulnerabilities, not to disrupt operations or cause data loss.

- **Full disclosure and remediation**: All discovered vulnerabilities must be promptly reported to the blue team for remediation.

By combining diverse skill sets, adhering to ethical principles, and employing realistic attack methodologies, your red team becomes a valuable asset in your ongoing quest for a more secure organization. Remember, the red team is not your enemy, but rather a critical partner in strengthening your defenses and staying ahead of evolving threats.

Blue team

Standing alongside the red team are the blue team members, defenders responsible for safeguarding your organization's systems and data. Their diverse skill sets and unwavering commitment are crucial for thwarting real-world attacks and continuously improving your security posture.

Blue team responsibilities encompass a wide range, including the following:

- **Security monitoring**: Continuously monitoring systems and networks for suspicious activity, often utilizing **security information and event management (SIEM)** tools.

- **Incident detection and response**: Analyzing potential security incidents, verifying threats, and initiating appropriate response measures to contain and mitigate damage.

- **Threat intelligence analysis**: Keeping abreast of evolving threats, attacker tactics, and vulnerabilities to proactively strengthen defenses.

- **Security automation**: Implementing automation tools to streamline detection and response processes, improving efficiency and accuracy.

In a nutshell, the blue team is responsible for building all the defensive controls for an organization and the red team tests the effectiveness by attempting to break them. By embracing these responsibilities, detection strategies, and a culture of continuous improvement, your blue team becomes a vigilant force, capable of proactively thwarting threats and ensuring your organization's security resilience. Remember, their dedication and expertise are invaluable assets in the constant battle against evolving cyber threats. Next, we will explore the key requirements for the success of red/blue teams.

Collaboration for success

The effectiveness of red/blue teams hinges not just on individual expertise, but on their seamless collaboration. Imagine not two opposing forces but partners united by a single goal: building a resilient security posture. This section sheds light on the fundamental pillars of successful collaboration: open communication, clear engagement rules, and a shared vision of security improvement:

- **Open communication**:

    - **Regular briefings and debriefings**: Foster information exchange through consistent briefings before red team engagements and comprehensive debriefings afterward.

    - **Transparency and trust**: Encourage open communication of successes and failures, fostering trust and allowing both teams to learn from each other.

    - **Dedicated communication channels**: Establish dedicated channels for real-time communication during red team engagements, ensuring clarity and swift response.

- **Clear engagement rules**:

    - **Clearly defined scope and objectives**: Outline the scope of the red team engagement, defining attack targets, acceptable techniques, and expected outcomes.

    - **Established escalation procedures**: Implement clear escalation procedures for unforeseen situations or potential incidents during the engagement.

    - **Real-time communication protocols**: Establish protocols for real-time communication between red and blue teams, ensuring both sides are aware of simulated attack progress.

- **Shared goal of security improvement**:

    - **Focus on the bigger picture**: Remind both teams that their ultimate goal is not winning a simulated battle, but collectively identifying and addressing vulnerabilities to improve overall security.

    - **Celebrate shared successes**: Recognize and celebrate successful detection and response efforts by the blue team, highlighting the value of both sides in improving security.

    - **Continuous learning and adaptation**: Encourage both teams to continuously learn from each other and adapt their approaches based on insights gained from engagements.

By fostering open communication, establishing clear engagement rules, and maintaining a shared focus on security improvement, red and blue teams can transcend their individual roles and function as a unified force, driving your organization toward a more resilient and secure future. Remember, collaboration is not just a means; it's the key to unlocking the full potential of your red and blue teams and solidifying your defenses against ever-evolving threats.

Conducting realistic red team engagements

The primary objective is to subject the organization's defenses to sophisticated and realistic attack scenarios, providing invaluable insights into vulnerabilities that might otherwise go unnoticed. By adopting a proactive adversarial approach, organizations can systematically assess and improve their resilience against advanced threats.

The success of realistic red team engagements hinges on meticulous planning, collaboration, and the integration of diverse skill sets. In this section, we will explore the key elements involved in crafting effective red team scenarios, from defining clear objectives to selecting appropriate attack vectors. Additionally, we'll delve into the importance of communication and coordination between the red and blue teams, ensuring a constructive and educational process. Through this exploration, you will gain practical insights into orchestrating red team engagements that not only challenge the organization's defenses but also serve as catalysts for continuous improvement in their security strategies.

Planning and scoping

The primary objective of the red team is to enhance the security posture of the organization rather than disrupt business operations and incur financial losses. In real-world scenarios, the true value of such exercises is realized when conducted in a production environment. This is primarily attributed to the fact that the most robust defense controls are typically deployed in production environments, which undergo rigorous monitoring. Consequently, meticulous planning and careful scoping of each engagement become a fundamental aspect to ensure the effectiveness of the exercise. Let's cover some of the important aspects of planning/scoping a red team exercise.

Defining the targets

With the goal of identifying potential targets for the exercise, here are some key things to keep in mind:

- **Alignment with business risks**: Prioritize targets based on their alignment with identified business risks and potential impact areas, such as critical infrastructure, sensitive data, or financial systems.

- **Simulated attacker mindset**: Consider your organization from the attacker's perspective, identifying attack vectors and high-value targets they might prioritize.

- **Phased approach**: Start with smaller, less critical systems for initial engagements, gradually progressing toward more complex and sensitive targets as the team gains experience.

Scoping the engagement

Once the target has been set, it is important to declare the boundaries, assumptions, and success criteria very clearly. Though a red team is simulating an external attacker, it is important to limit the scope of an engagement from a disaster recovery perspective. Let's explore some crucial aspects:

- **Clear objectives**: Clearly define the objectives of the engagement, such as identifying specific vulnerabilities, testing detection and response capabilities, or evaluating security controls.

- **Authorized attack methods**: Establish clear boundaries outlining the authorized attack methods, tools, and techniques the red team can utilize.
- **Exclusion zones**: Define any off-limits areas or systems to ensure the engagement remains safe, controlled, and focused on authorized targets.

Expected outcomes

Before beginning the engagement, it is very important to define exit criteria. This is to ensure red team engagements are finite and team members are not spending too much time on a very well-fortified system. It is also to ultimately aid in the process of enhancing security controls. Here are a few aspects to keep in mind:

- **Identified vulnerabilities**: Anticipate the discovery of vulnerabilities within the defined scope, empowering the blue team to prioritize patching and mitigation efforts.

> **Note**
>
> Prematurely sharing anticipated vulnerabilities can present challenges and may necessitate the involvement of an intermediary authority, such as the **Chief Information Security Officer** (**CISO**), to influence the blue team rather than engaging with them directly. To optimize the effectiveness of a red team exercise, it is advisable to steer clear of implementing custom and ineffective security controls, such as blocking a specific internal IP (which could be a red team member). Whether one is part of the blue or red team, it is crucial to recognize that the primary objective of the exercise is to benefit the organization. For CISOs, careful attention to the incentivization structure post-engagement becomes paramount.

- **Enhanced detection and response**: Expect the red team to challenge your blue team's detection and response capabilities, providing valuable insights for improvement.
- **Actionable recommendations**: Aim for comprehensive post-engagement reports with actionable recommendations for remediating vulnerabilities, strengthening controls, and improving overall security posture.

By meticulously planning and scoping your red team engagements, you set the stage for a valuable learning experience. Both teams gain valuable insights, allowing you to identify critical weaknesses, refine your defenses, and ultimately emerge with a more robust security posture. Remember, clear planning and defined expectations pave the way for successful Red Teaming, ensuring it becomes a cornerstone of your ongoing security improvement journey.

Simulating real-world attacks

Red teams don't just poke and probe; they unleash a simulated onslaught, mimicking the TTPs of real-world attackers. This section dives into the tools they employ, from exploiting vulnerabilities to leveraging social engineering, to creating a realistic and challenging training ground for your blue team.

A real-world persistent attacker is not just looking for unlocked doors into your organization, but also studying the ins and outs of your systems and exploiting human vulnerabilities. Red teams operate similarly, utilizing the following:

- **Reconnaissance**: Gathering information about your systems, network infrastructure, and personnel through **open source intelligence (OSINT)** and internal sources [6].

- **Vulnerability exploitation**: Employing known exploits and zero-day vulnerabilities to gain initial access to systems and networks.

- **Lateral movement**: Once inside, moving laterally across your network, escalating privileges, and seeking high-value targets.

- **Social engineering**: Utilizing psychological tactics such as phishing emails or pretext calls to trick users into revealing sensitive information or granting access. Some organizations may prefer to run periodic phishing campaigns, but it is often the entry point to systems.

Red teams also utilize deception techniques to test your detection and response capabilities:

- **Planting false flags**: Leaving behind misleading evidence to divert attention from their actual attack path. Security is an art of deception; much like real-world attackers, in-house red teams can employ deceptive techniques to bypass security gaps in your defense systems.

- **Denial-of-service attacks**: Overwhelming your systems with traffic to disrupt normal operations and impede response efforts. Especially if availability is one of the main concerns for your systems, it is important to ensure systems can detect sophisticated denial-of-service attacks.

By simulating these diverse attack methods, red teams provide invaluable insights into your security posture's weaknesses and blind spots. The blue team, forced to react and adapt, gains practical experience in identifying, containing, and mitigating real-world threats.

Measuring the impact

The true value of red teaming lies not just in simulating attacks, but in measuring their impact and gleaning actionable insights for security improvement. This section explores how to evaluate successful attacks, assess control effectiveness, and identify critical vulnerabilities exposed during red team engagements.

Dissecting a successful attack becomes crucial to understanding what went wrong. This can be done with the following:

- **Attack path analysis**: Trace the steps taken by the red team, understanding their initial entry points, exploited vulnerabilities, and lateral movement techniques.

- **Impact assessment**: Evaluate the potential damage caused by a successful attack, considering data loss, operational disruption, and reputational harm.

- **Control bypasses**: Analyze how the red team bypassed or disabled your security controls, identifying weaknesses in your overall defense strategy.

After getting a sufficient understanding of how the attack was executed, it is important to recognize both the existing controls that made this engagement harder and missing controls that could have made it impossible. Some key metrics to pay attention to are the following:

- **Effectiveness of detection and response**: Evaluate how well your blue team detected and responded to the simulated attack, measuring their speed, accuracy, and communication effectiveness.

- **Control activation and performance**: Analyze the performance of specific security controls triggered during the engagement, identifying any false positives, or missed detections.

- **Control gaps and overlaps**: Identify areas where controls were absent or ineffective, as well as potential redundancies or overlapping functionalities.

With the knowledge of the control gap and the vulnerability exploited, one last piece in this puzzle becomes measuring the impact and likelihood of the same vulnerability exploited by an external attacker. A follow-up exercise will be to understand the following:

- **Priority based on impact**: Focus on vulnerabilities exploited by the red team that could have led to the most significant consequences.

- **Likelihood of real-world exploitation**: Consider the likelihood of real-world attackers using similar vulnerabilities, prioritizing those posing the highest risk.

- **Remediation difficulty**: Evaluate the effort and resources required to patch or mitigate identified vulnerabilities, balancing risk reduction with feasibility.

Remember, red team engagements are not about assigning blame, but about learning and improvement. By effectively measuring the impact of red team engagements and translating findings into actionable steps, you transform simulated attacks into powerful catalysts for security improvement. This section provided a framework for analyzing red team results, emphasizing the importance of identifying high-impact vulnerabilities and prioritizing remediation efforts based on risk and feasibility. Feel free to adapt the content further to align with your specific security metrics and reporting procedures.

In the next section, we will develop a systematic approach to turn these findings into action items to drive up our organization's defensive resiliency.

Translating insights into actions

The primary objective of any red team engagement is to elevate the security standard with each iteration. Mere execution of exercises and identification of weaknesses in an organization's defense systems falls short of the ultimate goal. The true measure of success lies in leveraging these exercises as catalysts for continuous improvement, pushing the boundaries of security resilience higher in a progressive and iterative manner.

As we wrap up the exploration of red and blue teams within a security organization, our focus sharpens on turning the findings and insights into practical steps. This isn't just about spotting vulnerabilities; it's about taking those insights and actively making our defenses stronger. Imagine it as the moment when we don't just identify weaknesses but turn them into proactive strategies for better security. In this section, we'll delve into how red and blue teams can work together to transform their discoveries into concrete actions, helping organizations build resilient and adaptive security measures.

Analyzing results

Red team engagements generate a wealth of data, but the true value lies in transforming that data into actionable insights. Let's dive into the process of decomposing red team findings, identifying key weaknesses, and prioritizing remediation efforts to solidify your security posture.

Deconstructing the data

In a real-world attack, we find ourselves meticulously analyzing evidence to piece together the attacker's actions and identify vulnerabilities exploited. Red team analysis follows a similar approach. Let's highlight some of the deliverables:

- **Detailed reporting**: The red team provides a comprehensive report outlining their attack methods, exploited vulnerabilities, and achieved objectives.

- **Log analysis**: Analyze system logs, security event logs, and network traffic logs to understand the attacker's movements and identify specific events triggered during the engagement.

- **Blue Team debriefing**: Gather insights from the blue team, understanding their detection attempts, response actions, and any challenges encountered.

Comprehending these vital components of the puzzle is essential for addressing the identified gaps effectively. Let's take a further step down in looking at the exact findings.

Identifying key weaknesses

With the data deconstructed, it's time to pinpoint the critical vulnerabilities. We will further classify the findings into smaller categories:

- **High-impact exploits**: Prioritize vulnerabilities exploited by the red team that could have led to significant data breaches, system disruptions, or financial losses.

- **Control bypasses**: Analyze how the red team bypassed or disabled security controls, revealing weaknesses in your overall defense strategy.

- **Unforeseen attack vectors**: Identify any attack vectors or vulnerabilities the red team exploited that were not previously considered, expanding your threat awareness.

Each of these attack paths offers valuable insights for the blue team, with some illuminating aspects that can enhance our understanding of the current defense posture.

It is very important to make sure we don't get bogged down in analyzing every detail. But only focus on the high-impact findings and prioritize accordingly. It is often prudent to involve stakeholders from different departments, such as IT, legal, and risk management, to ensure comprehensive remediation plans.

By effectively analyzing red team results and translating findings into prioritized remediation efforts, you ensure the valuable insights gained from simulated attacks lead to tangible improvements in your security posture.

Refining defenses

Red team engagements unveil your security posture's strengths and weaknesses. Now we will explore how to leverage their findings to refine your defenses, encompassing patching vulnerabilities, enhancing detection capabilities, and improving response procedures:

- **Patching the findings**:

  - Focus on patching vulnerabilities deemed high-impact and easily exploitable, as identified through red team analysis.

  - Establish efficient patch management processes to ensure the timely deployment of security updates across your systems.

  - For unpatchable vulnerabilities, explore alternative mitigation strategies such as compensating controls or network segmentation.

- **Sharpening detection skills**:

  - Based on red team attack methods, update your detection rules to identify similar tactics and techniques employed by real-world attackers.

  - Utilize threat intelligence feeds to stay informed about emerging threats and incorporate **indicators of compromise (IOCs)** into your detection systems.

  - Implement deception techniques such as honeypots to lure attackers and gain valuable insights into their methods and tools.

- **Streamlining response**:

  - Analyze red team engagement results to refine your incident response playbooks, ensuring clear roles, responsibilities, and communication protocols.

  - Conduct regular incident response drills and tabletop exercises to simulate real-world scenarios and test the effectiveness of your response procedures.

  - Automate repetitive tasks such as containment and initial investigation to expedite response times and minimize damage.

These are foundational pointers to stimulate the contemplation process of solidifying the resilience of your organizational defense. By translating red team findings into targeted actions, you transform your security posture from reactive to proactive. Remember, defense is an ongoing effort. Let's now turn our focus toward continuous improvements.

Continuous improvements

Red teaming shouldn't be viewed as a one-time event, but rather a woven thread within the fabric of your ongoing security life cycle. This section emphasizes the importance of integrating red teaming into your security strategy and fostering a culture of proactive threat hunting for continuous improvement.

Initial red team engagements might be discouraging in some cases as they find more and more caveats in your defense design. However, integrating red teaming as part of your SDLC can be a game changer. Imagine before you launch a product publicly, a dedicated staff of security professionals vet the product, trying to tear it apart in all ways imaginable. In addition to hiring a red team, you can consider scaling your red team beyond your company doors. Bug bounty programs often offer lucrative sums for selected private groups of ethical hackers to test the security of an application before launch.

In conclusion, regular red team engagements provide invaluable insights into your evolving threat landscape and security weaknesses. Remember, security is not a destination, but a continuous journey. By embracing red teaming as a core part of your security strategy, you equip your team with the knowledge and skills required to stay ahead of cunning adversaries and navigate the ever-shifting threat landscape with confidence.

Targeted approach to controls and strategies

In the final leg of our exploration into uncovering weak points through an adversarial lens, we transition into the realm of a targeted approach for controls and strategies. Having conducted both organizational risk profiling and red team exercises, you've gained valuable insights into your security posture's strengths, weaknesses, and potential attack vectors. This section delves into how to effectively leverage these learnings to make targeted investments in your security controls and strategies, maximizing your return on security investment.

Building upon the foundations laid in the preceding sections on organizational risk profiling and red team exercises, we are poised to translate these insights into concrete and strategic defensive actions. This section unfolds in two parts, each directed toward honing our understanding and fortifying our defenses based on the dual perspectives obtained from risk assessment and simulated attacks.

The initial part focuses on distilling the outcomes of organizational risk profiling into actionable strategies. By categorizing and prioritizing potential threats based on the nature of the data, we aim to guide you in crafting targeted controls that align with their specific risk landscape. This deliberate approach ensures that defensive measures are not generic but finely tuned to the nuances of the organization's operations and potential adversaries.

The second part immerses us in the strategic aftermath of red team exercises. With a keen eye on the lessons learned from simulated attacks, we delve into formulating controls and strategies that directly counter the identified weaknesses. This integration of red team insights propels us beyond conventional risk-scoring models, enabling us to prioritize vulnerabilities according to their attractiveness to potential attackers. The amalgamation of these targeted approaches cultivates a DiD strategy that is not only robust but also dynamic, adapting to the ever-evolving threat landscape.

Leveraging risk profiling

In this chapter, we've taken a close look at how to identify specific threat vectors targeting our organization. Our goal was to create a framework to guide you through the process of theoretical approaches, helping align your security priorities with the most relevant risks. Now, in this section, we'll take what we've learned and use it to craft a DiD security strategy.

Leveraging your risk profile allows you to prioritize specific roads, investing strategically in controls that address your most critical vulnerabilities and high-impact threats. Let's move beyond the theory. You've identified your key assets, data types, and potential attack vectors through risk profiling. Now, let's translate that knowledge into actionable steps for customizing your security controls.

In the subsequent part of this section, we will formulate a strategic approach and design defensive controls tailored for widely used software with a user base of one billion. Let's delve into the key characteristics of this software:

- It stores basic user profile information and tracks user activity to provide personalized recommendations.
- It may store users' sensitive private information, such as credit card data, for ease of usage.
- It is a global software, run by billions of people daily.

Understanding the context is crucial in anticipating potential attacks on our product. Picture this scenario: our responsibility is to ensure the software safeguards user trust, prioritizes safety, and contributes to the overall security of both individual users and society at large. While this objective might resonate with many of you, there are specific considerations I'd like to highlight. Drawing from the insights gained in the preceding sections of this chapter, let's systematically analyze the potential consequences if an unauthorized entity gains access to our systems. In this examination, we'll concentrate on two distinct attacker groups:

- State sponsored
- Financially motivated

While financially motivated attackers might target user credit card information to extract money from victims of breached data, a state-sponsored attack group might just want to leverage the fact that the software is run by billions of users daily. In *Chapter 4*, we covered several TTPs and common patterns used by various attack groups. Connecting these dots, we can effectively focus our energy on parts of the system, not only from a technical defensive control perspective but also from a DiD point of view.

The 2017 Equifax breach exposed the personal data of millions due to a seemingly simple vulnerability – an unpatched Apache Struts application [7]. While patching might seem basic, understanding the potential impact of such a breach on a financial institution and the attacker's likely focus on exploiting unpatched systems make targeted investments in vulnerability management crucial. In the context of our hypothetical scenario, if vulnerabilities in a component of our system can expose user credit card information, we should consider patching them top priority.

Securing a system is undeniably a complex task, as vulnerabilities may exist in various parts. While the recommendation to patch software seems straightforward in hindsight, the real challenge often lies in accurately identifying all dependencies. This is precisely where the concept of DiD becomes invaluable. Particularly around the critical assets representing the crown jewels of your organizational data, conduct adversarial simulations. By envisioning potential compromises within different layers of the system, you can strategically apply resilient controls, fortifying your defenses comprehensively.

By leveraging your risk profile as a roadmap, you move beyond generic security and make informed decisions about where to invest your resources. This targeted approach ensures your security controls provide maximum protection for what matters most, ultimately improving your overall security posture and resilience. Now let's look at the same problem at a later stage of the process.

Building on red team exercises

Your red team exercises have shed light on your security posture's blind spots and the attacker's cunning tactics. Now, it's time to translate those insights into strategic action, selecting and implementing effective security controls that truly fortify your defenses.

Ideally, we will now be familiar with the overall risks our organization faces. In addition, we have received insights from our red team about a recent engagement they carried out in attempting to install a backdoor into our application. Imagine the impact of a successful attack; the attacker would basically have access to billions of user devices across the globe! For this discussion, let's assume they came back with two different ways of doing that:

- By compromising a well-secured internal server and subsequently injecting their crafted payload to install a backdoor.

- By exploiting the authentication/validation logic in the credit card upload part of the application, thus able to deliver the payload directly.

Given the class of attackers we are dealing with are both high-severity issues within our systems, if we had the resources to fix only one, which one would you pick first? Think beyond individual controls. Consider how controls work together to create a layered DiD strategy. A malicious insider might be able to exploit the first vulnerability, but your organization's stringent code review policies might help here. Consider that red team members were able to bypass the authentication/validation logic, which is likely reused in other places, and it is exposed externally. Don't get me wrong, both findings lead to full remote code execution, so either would be okay for an attacker, but the second finding comes with no ifs and buts.

Strategic considerations for control selection

Here are some common strategies for choosing the right control:

- Never over-complicate your defenses.

- Ensure the chosen controls directly address the threats and risks identified in your risk profile and red team findings.

- Balance the effectiveness of a control with its implementation and maintenance costs. Prioritize controls that offer high value relative to their resource demands.

- Consider how well the control integrates with your existing security infrastructure and tools. Aim for seamless interoperability for efficient management and analysis.

- Choose controls that can be scaled to accommodate future growth and are easily maintained within your technical expertise and resource constraints.

By strategically selecting and implementing controls based on your red team findings and risk priorities, you move beyond generic solutions and build a targeted defense that effectively addresses your unique security needs. This ensures your resources are invested wisely, maximizing your return on security investment, and ultimately creating a more resilient security posture.

Summary

Building upon the attacker insights gained in *Chapter 4*, this chapter translated that knowledge into action via practical frameworks. We began by crafting a data-driven organizational risk profile, utilizing techniques such as adversarial simulation to identify weaknesses. This informs risk prioritization based on the attacker's perspective, ensuring focused mitigation efforts. After that, we delved into setting up a successful security organization, exploring the key factors that enable red and blue teams to experience optimal collaboration. Finally, to bridge theory and practice, we walked through a hypothetical scenario showcasing real-world application of the concepts covered. The tools and frameworks introduced in this chapter will augment the ability to identify the unique threat landscape for your organization and aid in building a robust layered security strategy.

In the next chapter, we will progress a step further by meticulously examining prevalent attack patterns observed in the real world over recent years. The objective is to construct a mental map rooted in our collective comprehension of the attacker's perspective, ultimately enhancing our defensive methodologies for a more strategic advantage.

Key takeaways

- An attacker perspective, even if not perfect, can fundamentally change the way you look at systems.

- Data is almost always at the center of persistent exploit attempts in your organization. Reducing the data footprint is the single most effective way to reduce your attack surface.

- If you don't know your data, you can't protect it. Establishing and sustaining a thorough data inventory is of utmost importance.

- You can make an educated guess on who's going to attack your systems if you have clarity on what kind of data your organization holds.

- Simulating damage often resolves the debate around a system's worth.

- Red and blue teams might have short-term conflicting goals, but both are trying to solidify the organization's security posture.

- A successful red team engagement always results in easy and repeatable proofs of concept.

Congratulations on getting this far! Pat yourself on the back. Good job!

Further reading

To learn more about the topics that were covered in this chapter, take a look at the following resources:

- [1] The Most Popular Computer Software, Ranked: `https://strawpoll.com/most-popular-computer-software`

- [2] Data-hungry apps: These are the worst for your privacy: `https://www.komando.com/security-privacy/data-hungry-apps/852537/`

- [3] Netflix Chaos Monkey: `https://netflix.github.io/chaosmonkey/`

- [4] MITRE ATT&CK, cyber kill chain, and purple teams: `https://www.sans.org/blog/cyber-kill-chain-mitre-attack-purple-team/`

- [5] Google AI Red Teams: `https://blog.google/technology/safety-security/googles-ai-red-team-the-ethical-hackers-making-ai-safer/`

- [6] OSINT framework: `https://osintframework.com/`

- [7] Equifax data breach FAQ: `https://www.csoonline.com/article/567833/equifax-data-breach-faq-what-happened-who-was-affected-what-was-the-impact.html`

Mapping Attack Vectors and Gaining an Edge

Up to this point, we've dissected the attacker's playbook, exposing their techniques and pinpointing weaknesses within our own defenses. It's time to level up. In this chapter, we'll analyze specific attacks, tracing them back to the motivations and capabilities of different attacker profiles. This intelligence isn't just about understanding; it's about empowerment.

This isn't just about knowledge for knowledge's sake. By grasping who is most likely to come after you and how they choose to operate, you'll gain an invaluable edge. With this insight, you won't waste time or resources on generic security. Instead, you'll target your patching and defenses with laser precision, making attackers think twice.

We'll explore how to analyze attack patterns, understand the varying capabilities of different threat actors, and tailor your defenses accordingly. The goal isn't simply to fight back; it's to create a security posture so robust that potential attackers decide you're simply not worth the effort.

In this chapter, we're going to cover the following main topics:

- The anatomy of common attack vectors
- Linking attack vectors to attacker profiles
- Building proactive defensive programs

Let's get started!

The anatomy of common attack vectors

In this initial segment, we will delve into prevalent attack vectors that have been increasingly leveraged to impact organizations on a large scale. By dissecting various attack patterns, we aim to familiarize ourselves with the evolving threat landscape. Furthermore, an examination of these attack patterns against our own organizations often unveils a clandestine cohort of adversaries endeavoring to breach our systems. Building upon our earlier exploration in *Chapter 4*, where we extensively analyzed several attacker profiles and their primary targets, and *Chapter 5*, which elucidated methodologies for identifying weaknesses in our systems through an adversarial lens, our current focus shifts to the ongoing monitoring of attacks targeting our organization and similar entities. The objective is to discern recent attack patterns and fortify our defenses accordingly to thwart such incursions.

Next, we'll dissect common attack types using real-world case studies of significant exploits. This analysis will provide a tactical understanding of attacker methodologies.

Network exploits

Network exploits encompass a broad range of attacks that target vulnerabilities in systems accessible over a network. This includes weaknesses in network devices (e.g., firewalls, routers, etc.) and vulnerabilities within software that allow attackers to send malicious packets. Whenever an attacker leverages network connectivity to compromise a remote system, the attack is broadly classified as a network exploit.

Network exploits take advantage of flaws in the systems and protocols that connect our devices. These attacks can target anything from vulnerable routers and firewalls to specific software running on servers or individual machines. Understanding network exploits is crucial, as they often provide the initial foothold that attackers need to breach an organization.

Common types of network exploits

The following are some common attack patterns classified under this exploit group:

- **Vulnerability exploits**: These target known weaknesses (e.g., unpatched software, misconfigurations, etc.) with targeted malicious code.
- **Zero-day exploits**: These target previously unknown vulnerabilities for which no patch or fix exists, making them highly dangerous.
- **Denial-of-service (DoS) attacks**: These overwhelm systems with traffic, rendering them inaccessible to legitimate users [1].
- **Person-in-the-middle attacks**: Attackers intercept and manipulate traffic between two systems, either eavesdropping or altering data.

Let's turn our attention to a real-world example of a successful network exploit.

Real-world example – The Log4j vulnerability (2021)

In December 2021, the Log4j vulnerability rocked the tech world [2]. Log4j is a popular logging library used in many Java applications. Attackers discovered they could send specially crafted text through various input channels (e.g., website forms) to the application using Log4j. This could trigger remote code execution on the vulnerable server, opening the doors to a full system takeover.

The Log4j vulnerability, known as CVE-2021-44228 [3] and more, resided in the way this widely used logging library processes text. Attackers could embed malicious code within logged messages (via user input, network connections, etc.). When processed by Log4j, this code could trigger remote code execution on the target system.

The impact was so vast due to the extensive use of Log4j across both open source and commercial software, running on everything from simple devices to sophisticated web servers. On the other hand, the attack was deceptively simple to perform. This amplified the risk, allowing even less skilled attackers to wreak havoc.

While virtually any organization using affected software was at risk, the consequences were acutely felt by the following:

- **Enterprise and cloud services**: Companies with large software stacks, often including numerous Java applications reliant on Log4j. The sheer number of vulnerable systems demanded intensive patching efforts.

- **Web-facing organizations**: Traditionally, web applications are commonly built on Java. Those with user input fields (e.g., forms, file uploads, etc.) were prime targets, increasing the likelihood of successful exploitation.

The Log4j vulnerability is a prime example of zero-day being exploited by threat actors. It underscored the dangers of vulnerabilities deeply embedded within widely used software libraries and the imperative of vigilant patching.

Web application attacks

Web applications form the front line of interaction for many organizations. Sadly, this also makes them prime targets for attackers. Application attacks focus on vulnerabilities within the code and logic of the application rather than network-level or system-level weaknesses.

The list of web application attacks demands a book of its own and we will not go into detail in this book; however, let's take a high-level overview for our understanding of such attacks and associated attacker profiles.

Common types of application attacks

The following are some common attack patterns classified under this exploit group:

- **Cross-site scripting** (**XSS**): Injecting malicious JavaScript into a web page, executed by unsuspecting users' browsers. Stealing session cookies, redirecting users, or altering web page content are common goals of XSS attacks. However, these attacks are often utilized by attackers to gain an initial foothold of the application.

- **SQL injection**: Manipulating database queries to extract sensitive data, modify records, or even take control of the underlying database.

- **Command injection**: Tricking the application into executing unintended system commands, allowing attackers to compromise the web server itself.

- **Buffer overflows**: Sending overly long input that overwrites memory areas, enabling attackers to execute arbitrary code on the server.

Let's turn our attention to a real-world example of a successful application exploit.

Real-world example – The Equifax breach (2017)

The Equifax data breach stands as a catastrophic example of the consequences of an application-level exploit [4]. Attackers targeted a vulnerability in the Apache Struts framework used by a web application. This flaw allowed them to execute remote code on Equifax servers, ultimately leaking the **personally identifiable information** (**PII**) of millions of individuals.

Equifax used the Apache Struts 2 framework for its web-facing consumer dispute portal. A critical remote code execution vulnerability in Struts 2 (CVE-2017-5638) [5] was discovered and publicly announced in March 2017.

Equifax failed to adequately patch this vulnerability in a timely manner, leaving systems exposed. Between May and July 2017, attackers discovered and exploited the Struts 2 flaw on Equifax's system, gaining access to the sensitive data of over 147 million people.

The massive Equifax data breach highlights the following organizations as being most vulnerable to similar attacks:

- **Organizations with extensive consumer data**: Businesses that store PII – such as financial institutions, healthcare providers, and retailers – are high-value targets for breaches such as that of Equifax.

- **Organizations using vulnerable web frameworks**: For companies that rely heavily on web application frameworks (Struts, Spring, etc.), without rigorous patch management and secure coding practices, vulnerabilities in these frameworks expose the organization to significant risk.

- **Organizations with large attack surfaces**: Complex web applications with many functionalities and user input points provide ample opportunities for attackers to find and exploit weaknesses.

This is a prime example of a large-scale data exfiltration exploit launched by exploiting an application vulnerability. The Equifax breach exposed the severe consequences of failing to proactively address web application vulnerabilities.

Social engineering

Social engineering attacks sidestep technical vulnerabilities. Instead, they rely on psychological manipulation to trick people into performing actions detrimental to security. While less overtly "technical" than some other attacks, they're insidious and effective, and sometimes, the precursor to technical compromise.

Despite their less overtly technical nature, social engineering attacks are a persistent component of an attacker's toolkit. These attacks are highly effective for initial reconnaissance, as seemingly innocuous information (such as password policies or software versions) can be invaluable to a dedicated adversary.

Common types of social engineering attacks

The following are some common attack patterns classified under this exploit group:

- **Phishing**: Fraudulent emails or messages designed to obtain sensitive information (logins, financials, etc.) or encourage interaction with malware attachments or links.

- **Pretexting**: Crafting a believable fabricated scenario (e.g., impersonating IT support or an authority figure) to extract information or get the victim to perform an action [6].

- **Baiting**: Enticing victims with something desirable (e.g., a free USB drive) loaded with malware, or tricking them into accessing infected online resources.

- **Tailgating**: Physically following an authorized person into a restricted area by blending in (e.g., holding a coffee or appearing to be talking on the phone).

> **Note**
> Social engineering attacks often evolve along with current events and prey on emotions such as fear and urgency. Maintaining up-to-date awareness of trends is vital to defense.

Let's turn our attention to a real-world example of a successful social engineering attack.

Real-world example – The Twitter hack (2020)

In a high-profile attack, hackers used social engineering tactics to take over the Twitter accounts of famous individuals (such as Elon Musk, Bill Gates, etc.) to promote a Bitcoin scam [7]. Attackers gained access to internal company tools via phone-spear-phishing attacks on Twitter employees, showcasing how social engineering can undermine even a tech giant.

The attack primarily relied on phone spear phishing to target Twitter employees with access to internal support tools. Attackers likely researched their targets using public information to refine their approach and seem credible.

Through manipulation, attackers convinced employees to provide access credentials and even to perform specific actions within Twitter's internal systems. Having compromised powerful administrative tools, attackers posted the infamous Bitcoin scam tweets from verified, high-profile accounts.

Organizations of all sizes can be hurt by social engineering attacks, but the Twitter breach showed just how severe the risks are for the following:

- **Social media/high-profile tech platforms**: Reputational damage and the ability to spread misinformation on such platforms amplify the potential impact of breaches such as this.

- **Organizations with high-value accounts**: Businesses whose customers, partners, or executives have a large following can see those channels weaponized for financial scams or even more damaging content.

The Twitter attack in 2020 taught us that even robust technical defenses can be negated if employees are tricked or coerced. Security training that reinforces recognizing red flags in unsolicited contact (digital or physical) is vital. Even tech-savvy companies aren't immune when staff aren't trained in spotting social engineering schemes. Regular exercises and clear reporting methods are vital. Also, limiting administrative tool privileges and requiring multi-step verification for any sensitive action helps contain damage, even if initial credentials are compromised.

If you are interested in learning more about social engineering attacks, my personal favorite is following the DEFCON Social Engineering Village [8] where participants make live calls to attempt social engineering on real people/companies.

Insider threats

Insider threats come from those with legitimate access to your systems – current or former employees, contractors, and partners. This unique position provides them with knowledge and privileges that external attackers typically struggle to obtain.

You might have heard that *security through obscurity* does not work. This attack class puts the spotlight on why it is important for high-value companies to account for insider risk.

> **Note**
>
> Imagine you're designing a distributed system. You have set a communication pattern between microservices. A predefined pattern in the request determines whether it will be allowed or denied. The security model relies on the secrecy of this request pattern. This is a classic example of security through obscurity. Your assumption here is that the system is secure because we're not going to reveal this pattern externally.

Types of insider threats

We will explore different types of insider risks and their motivations:

- **Malicious insider**: Actively working to harm the organization by stealing data, sabotaging systems, or leaking information. Their motivation can be financial gain, revenge, and even ideological reasons.

- **Negligent insider**: Inadvertently putting the organization at risk through errors, poor security practices (weak passwords, falling for phishing, etc.), or not following protocols. They lack malicious intent, but the consequences can be equally severe.

- **Compromised insider**: An unwitting pawn whose account is hijacked by external attackers (e.g., through phishing), then leveraged for malicious purposes.

Let's turn our attention to a real-world example of a successful insider attack in action.

Real-world example – The Uber data breach coverup (2016)

Uber suffered a significant data breach where the information of 57 million users and drivers was accessed. Even more damaging, Uber's former **chief security officer** (**CSO**) and team paid the attackers to delete the stolen data and attempted to disguise the breach as a "bug bounty" program. This exemplifies malicious insider actions and the devastating consequences of attempting to conceal them. [9]

An initial foothold was gained by accessing private GitHub repositories used by Uber developers, where authentication credentials for an Uber data store on **Amazon Web Services** (**AWS**) were exposed. This enabled attackers to download the private information (usernames, email addresses, phone numbers, driver's license details, etc.) of millions.

Instead of disclosing the breach, as required by law, the Uber CSO initiated a payment of $100,000 in Bitcoin to attackers with demands they destroy the data. The payment was disguised internally as a "bug bounty" payout; further attempts were made to hide the involvement of outside attackers.

It wasn't until a year later, under new leadership, that the breach was disclosed. Legal fines, class-action lawsuits, and massive brand damage resulted. The former CSO faces charges of obstruction of justice for the coverup attempt.

The Uber incident underscored the insider threat risks that are particularly concerning to the following:

- **Data-centric companies**: Organizations operating as data aggregators (ride-sharing apps, customer loyalty programs, digital platforms, etc.) often possess highly sensitive user databases that are immensely valuable to cyber criminals.

- **Companies with extreme reach**: Major multinational companies with reach beyond billions of users across the world are often targeted by nation-state attackers disguising themselves as innocuous insiders.

- **Companies with poor security hygiene**: Mishandling of login credentials, limited visibility into systems, and a lack of security policies exacerbate the potential for malicious insiders to exploit them.

- **Rapidly growing businesses**: Prioritizing business development over thorough security practices creates opportunities for disgruntled employees or internal oversight gaps to facilitate attacks.

- **Highly competitive industries**: Competitors might actively attempt to recruit key employees with system access to obtain trade secrets or disrupt operations.

> **Note**
>
> Insider threats are often more complex to defend against than external attacks due to the pre-existing level of trust. Proactive security culture and technical measures play a crucial role in mitigation. Defense in depth is the key; we will come back to how these attacks could have been prevented with a simple layered strategy in the next chapter.

Malicious actions of this scale show how internal individuals can directly lead to major reputational and financial damage. Attempting to conceal a breach often makes the situation far worse; having clear incident response procedures and commitment to transparency are vital.

Supply chain attacks

Supply chain attacks target less secure elements within a network of suppliers, vendors, and third-party software that support an organization's operations. The primary target might be your business, but attackers indirectly gain access by infiltrating and exploiting trust in your digital "supply chain."

Over the past few years, supply chain exploits have been gaining popularity because of their potential exponential impact. This is attributed to the fact that our open source software ecosystem is getting very mature, causing in-house development of commonly used software components inefficient, and ultimately leading to a rise in the usage of these open frameworks across large organizations. Thus, exploiting vulnerabilities in well-known packages can yield exponential returns for an attacker.

Common types of supply chain attacks

The following are some common compromises classified under this exploit group:

- **Software compromises**: Injecting malicious code into trusted software updates or even open source components that an organization then unknowingly integrates into its own systems.

- **Hardware tampering**: Interfering with the production process of devices (routers, laptops, point-of-sale systems, etc.) to introduce backdoors or vulnerabilities, undermining security before they are even deployed.

- **Third-party service providers**: Leveraging the access a partner organization might have to a network. An example is breaching a managed service provider to then "pivot" access and deploy attacks against its clients.

Let's turn our attention to a real-world example of a successful supply chain attack in action.

Real-world example – The SolarWinds attack (2020)

Considered one of the most sophisticated supply chain attacks, it started with attackers compromising SolarWinds' Orion network monitoring software [10]. Tainted software updates were then rolled out to 18,000+ SolarWinds customers.

Attackers could leverage this backdoor for nearly undetectable remote access, data theft, and additional malware deployment within target organizations, including major tech companies and government agencies.

Attributed to Russian state-sponsored actors, the attack exhibited extreme patience and careful operational security. Attackers infiltrated SolarWinds' build environment, injecting malicious code into Orion updates; this created a seemingly legitimate backdoor into thousands of target networks.

After gaining access, attackers often remained dormant for extended periods, mimicking normal network activity for maximum evasion. The focus seemed to be on long-term espionage rather than quick disruption. Over 18,000 SolarWinds customers received the tainted updates. From there, attackers selectively escalated attacks against high-value targets. This included US government agencies, cybersecurity firms, and Fortune 500 companies.

The SolarWinds attack starkly illuminated the potential for supply chain attacks to cause widespread damage. Organizations at most risk were the following:

- **Government agencies**: Access to sensitive government networks puts national security at risk. The SolarWinds attack drove an immediate focus on protecting government systems against supply chain attacks.

- **Large technology companies**: The breach allowed attackers to penetrate major industry players, causing ripple effects as attacks potentially pivoted to those companies' clients.

- **Critical infrastructure**: Some targeted organizations played key roles in power infrastructure and utilities – even brief disruptions could have a significant impact.

SolarWinds had an established software development practice, illustrating that sophisticated organizations can still fall victim to well-executed supply chain attacks. When attackers gain footholds in organizations that are central to operations in many industries, the ripple effects are widespread and long-lasting.

Physical attacks

Physical attacks involve direct attempts to gain unauthorized access to facilities and hardware and circumvent physical security measures. They may seem less sophisticated than complex malware, but the risk they pose shouldn't be underestimated. Let's quickly explore this final attack type for this section.

Common types of physical attacks

The following are some common attacks classified under this exploit group:

- **Theft of devices**: Laptops, smartphones, or even data servers being forcibly stolen, either due to targeted efforts or simple opportunism. Loss of hardware means potential access to unencrypted data, and attackers leveraging those devices to further penetrate internal networks.

- **Tailgating and social engineering**: Following staff into secured areas, pretending to be a repair person, or otherwise using manipulation to bypass physical security checks.

- **Tampering with hardware**: Installing keyloggers, hardware skimmers (often on point-of-sale systems), or deliberately damaging equipment to cause disruption.

- **Exploiting external connections**: Unauthorized access to poorly secured network ports or backup data facilities, or tampering with network cabling infrastructure.

Let's turn our attention to a real-world example of a successful physical attack in action.

Real-world example – Data center theft

While exact cases are often kept private, data center heists for hardware are a serious risk. They involve forcibly breaching server facilities, and sometimes, dismantling and making off with large volumes of servers containing vast amounts of valuable data. Such attacks are less about the value of the hardware itself, and more about obtaining the valuable information hosted on them.

> **Note**
> Physical attacks often are one step in a broader attack chain. A stolen device might offer initial entry, allowing later deployment of sophisticated digital attacks.

Remember that robust physical security measures (multi-factor access, surveillance systems, tamper-evident seals, etc.) in conjunction with digital protection create a resilient posture.

Now that we have explored various attack classes and their corresponding exploits to gain insights into common organizational targets and their impacts, let us now shift our focus to leveraging this information to establish connections between these attack vectors and our organization, as well as our targeted attacker profile.

Linking attack vectors to attacker profiles

In the *Threat actors and their motivations* section in *Chapter 1*, we described different types of attacker profiles and their primary motivations behind launching attacks. Later, in *Chapters 4* and *5*, we went over them in more detail to cover common techniques employed by different classes of attackers. We also discussed attacker profiles based on an organization's data assets.

In the earlier section, we took a tour of common attack patterns through recent, large-scale exploits. Let's now synthesize this knowledge to proactively detect attacks in real time. By combining threat intelligence, proactive monitoring, deceptive tactics such as honeypots, and analysis of key profiling indicators (sophistication, targets, timing, persistence, etc.), we can continuously uncover hidden adversaries targeting our systems.

Before jumping into the content, let's recap our learnings about types of threat actors from previous chapters using *Figure 6.1*:

Attacker Type	Description
Script Kiddies	Less sophisticated, often using pre-made tools
Organized Crime	Financially motivated, focus on data theft or ransom
Nation - State Actors	Espionage, critical infrastructure disruption
Hacktivists	Attacks driven by ideology

Figure 6.1 – Types of attackers

With the attacker types and motivations refreshed in our memory, we will now begin our two-stage journey to linking attack vectors to attacker profiles.

Defensive information gathering

Just as attackers perform extensive enumeration of our systems and environments, their defensive counterparts can equally perform information-gathering campaigns to gain invaluable information by accessing current threat landscapes, attacker profiles, and patterns targeting similar organizations to provide invaluable insights. Remember, cybersecurity is an ongoing, iterative battle. Understanding your adversary is essential for a shift to modern security.

Traditionally, security has been more of a reactive approach; however, to achieve real security milestones, we need to take more proactive actions. We will cover this in more detail in the next section by formulizing building proactive defense programs.

Now, let's explore common sources of information to get insights into ongoing attacks that might impact your organization.

Threat intelligence feeds

Threat intelligence feeds provide continuous streams of curated data about potential and ongoing cyberattacks. This data often includes **indicators of compromise (IoCs)** such as malicious IP addresses, malware hashes, domain names associated with attacks, and attacker **tactics, techniques, and procedures** (TTPs). By ingesting and analyzing this threat intelligence, security teams gain a real-time view of the threat landscape, allowing them to detect and block malicious activity before it compromises their systems.

Here's how threat intelligence feeds provide a proactive advantage over attackers:

- **Early warning system**: Feeds expose emerging threats and attack patterns as they evolve. This gives defenders time to patch vulnerabilities, adjust network configurations, and update defensive signatures before widespread exploitation occurs.

- **Threat contextualization**: While isolated alerts are useful, feeds often include context on specific attackers, their preferred targets, and their methodologies. This informs decisions on fine-tuning security controls and prioritizing potential threat responses.

Choosing which feed to use for your organization will depend on the industry, technology stack, relevance, and integrations, among other factors, but here are some prominent live threat intelligence feeds:

- **Open source feeds**:

  - **Abuse.ch**: A well-respected Swiss project specializing in malware, botnet tracking, and phishing-related threats. It offers feeds such as the Feodo Tracker (botnet IPs), SSL Denylist (compromised certificates), and URLhaus (recent malicious URLs) (`https://abuse.ch/`).

  - **Open Threat Exchange (OTX)**: AlienVault's community-driven feed with a massive pulse database. Users share IoCs, analysis, and insights into new threats (`https://otx.alienvault.com/`).

  - **Malware Information Sharing Platform (MISP)**: Promotes open sharing of indicators and analysis tools within the security community. It includes both free and subscription-based feeds (`https://www.misp-project.org/`).

- **Commercial feeds**:

  - **Recorded Future**: Offers real-time context around a range of threats, including IP reputations, domain threat assessments, and vulnerability intelligence (`https://www.recordedfuture.com/`).

  - **CrowdStrike Falcon X**: Leverages real-time endpoint data from CrowdStrike sensors, identifying malware and suspicious behaviors and linking them to targeted attacks (`https://www.crowdstrike.com/`).

You might want to use a combination of these feeds to avoid creating blind spots, but choosing which one applies to your organization might require more discussion than choosing from this list.

Traffic and pattern monitoring

Traffic and pattern monitoring involve analyzing the data flowing into, out of, and within your network. By establishing baselines of "normal" network activity, security teams can quickly spot anomalies that might indicate an active attack. This offers real-time visibility into malicious behavior, often during the very early stages of an attack. Analyzing denied traffic at security edge devices (WAFs, firewalls, etc.) provides valuable insights into the types of attacks targeting your organization. Remember that defensive controls that are effective today may not withstand evolving threats. Continuous analysis of blocked attempts is essential as attackers will tirelessly refine their strategies.

Honeypots

Honeypots are deliberately designed to appear as vulnerable systems or attractive targets, enticing attackers to interact with them. This creates a controlled environment where security teams can safely monitor and analyze attackers' behaviors, methods, and tools. Unlike production systems containing valuable data, there's a minimal risk of damage when an attacker compromises a honeypot.

> **Important note**
>
> The effective deployment of honeypots requires careful configuration to make them appear authentic to attackers. Depending on the complexity, they can range from low-interaction (simulating only basic services) to high-interaction (emulating full operating systems) to glean maximum data.

Honeypots have been very effective in "hacking the hacker," gathering intelligence on attackers interested in your organization by closely monitoring their activity in a contained environment under your control.

Having examined techniques for intelligence gathering on active attacks, let's now focus on using this data to establish threat actor profiles.

Key profiling indicators

Real-time observation of attacks not only provides current situational awareness but also illuminates shifts in the threat landscape. While it's understood that we cannot anticipate or prevent every potential attack against our organizations, real-time monitoring offers valuable insights into the profiles of attackers seeking to breach our defenses. This information enables us to extrapolate and apply our understanding of TTPs, as discussed in *Chapter 4*, to anticipate future attacks that align with specific attacker profiles. In this section, we will examine several key indicators that can aid in profiling the attackers targeting our organizations.

Sophistication of methods

The complexity and customization of an attacker's tools and techniques provide clues about their experience and level of resources. Analyzing attack sophistication helps you to gauge the skills you're up against.

The following examples will serve to reinforce our comprehension:

- **Simple exploits**: Script kiddies and opportunistic attackers often rely on widely available tools and basic attacks.

- **Targeted customization**: Spear phishing with well-researched lures, or attack chains tailored to your environment, suggest a more skilled and resourceful adversary.

- **Zero-day exploits**: Utilizing previously unknown vulnerabilities indicates a highly sophisticated actor with extensive technical capabilities or financial resources.

Targets

Examining the industries impacted, the specific data sought, and the geographic focus of attacks helps us discern whether adversaries are financially motivated, driven by espionage, or aiming for disruption. Understanding what attackers deem valuable exposes their intentions.

Let's dive into target patterns that can unveil underlying attacker profiles:

- **Industry**: Are attackers targeting a specific sector (finance, healthcare, energy, etc.)? This hints at their motivations and knowledge base.

- **Data type**: What data seems to be their target (customer records, intellectual property, system disruption, etc.)? This further refines potential attacker profiles.

- **Geographic focus**: Are attacks centered on one region, or spread globally? This helps pinpoint whether you're facing opportunistic threats or actors focused on a certain locale due to business rivalries or geopolitical motives.

Timing and persistence

Attack frequency, patterns in timing, and an attacker's determination reveal whether they are opportunistic or engaged in carefully orchestrated, long-term campaigns. Timing and persistence help define whether you're up against a fleeting nuisance or a dedicated opponent.

Here, we will examine various indicators that aid in assessing attacker profiles:

- **Opportunistic attacks**: Exploiting a widely reported vulnerability quickly suggests less-dedicated actors seeking easy gains.

- **Long-term campaigns**: Persistent, low-level probes over an extended period indicate high-value targeting by patient, stealthy attackers.

- **Timing cues**: Does attack activity mirror certain business cycles or geopolitical events? This can connect a seemingly isolated incident with a larger attacker agenda.

Let's formalize our understanding by putting all the pieces of puzzles we've learned about so far in *Chapters 4*, *5*, and *6* and looking at the bigger picture, as follows:

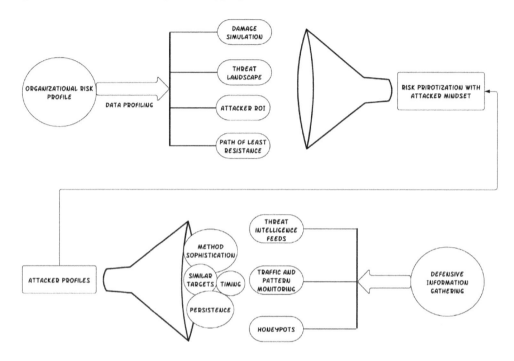

Figure 6.2 – Overall risk management with an attacker mindset

While paying attention to attacks in real time and mapping them back to attacker profiles can be truly valuable, it is important to note that profile indicators rarely give a 100% conclusive profile. Instead, they form a probabilistic model. Indicators might overlap between different types of attackers, but careful analysis enables strategic prioritization of defense resources.

For example, combined observations of advanced zero-day use, an interest in sensitive R&D data, and persistent activity suggest a well-funded nation-state actor or a competitor conducting economic espionage.

Building proactive defensive programs

Now that you're equipped with insights into attack anatomy and attacker profiles, this final section will provide a brief overview of the key characteristics of robust defensive programs, laying the groundwork for more detailed exploration in the next chapter. It's evident that in the face of an ever-evolving and increasingly sophisticated threat landscape, reactive security measures are inadequate. Therefore, a defender's toolkit must incorporate substantial proactive measures involving a vigilant assessment from an adversarial perspective and the implementation of counteroffensive strategies.

By "counteroffensive," I don't mean launching attacks against adversaries, but rather engaging with them more proactively through mechanisms such as threat intelligence and honeypots, as discussed earlier in the *Defensive information gathering* section.

Our analysis of attacks, from common techniques to specific threat actors targeting organizations such as yours, offers crucial lessons for strengthening your security posture. While the next chapter will dive into technical implementation, let's first establish the core principles guiding a proactive defense model:

- **Prioritize target vulnerabilities**: Think back to attacks such as Log4j and SolarWinds. Prioritize ruthless patching and software supply chain vetting. Vulnerability management isn't just a checklist – it's an urgent race against attackers.

- **Attackers aren't one-size-fits-all**: Threat profiling gives context. Financially motivated groups might be deterred by strong encryption and access control, even if some perimeter weaknesses exist.

- **Humans are a vulnerability**: Social engineering thrives in many incidents (see the Twitter hack). Regular security training tailored to employee roles is as critical as firewalls. Focus on phishing recognition and secure reporting for suspicious activity.

- **Assume it's already inside**: Modern threats often silently breach environments and then lie dormant. Proactive threat hunting (even within "trusted" software) becomes necessary. Consider network behavioral anomaly detection tools to support this.

- **Intelligence isn't static**: Threat feeds give you real-time attack data. Integrate these feeds with existing defenses to dynamically block emerging threats instead of manually updating rules after the fact.

Everything elucidates that the key concept is **defense in depth**. No single technology is a silver bullet. A proactive program incorporates multiple overlapping layers, from securing external-facing devices to hardening endpoints, and from monitoring the network to training your workforce. This makes even successful attacks less likely to gain the foothold needed to cause serious damage.

Summary

In this chapter, we dove headfirst into the world of cyberattacks, dissecting techniques, targets, and the driving forces behind various threat actors. Our analysis began broadly, looking at widespread attack patterns and how seemingly unrelated vulnerabilities (such as those revealed in Log4j) can trigger massive breaches. It's clear that staying attuned to the ever-shifting threat landscape is critical for organizations of all sizes.

Next, we took a step back, connecting those attacks to attacker profiles. Moving beyond the "hacker" stereotype, we learned that the sophistication of tools, the choice of targets, and the persistence of attacks provide vital clues. Whether facing opportunistic adversaries or highly skilled and targeted campaigns, these profiles illuminate a threat actor's likely goals and the level of defense required.

This chapter laid the groundwork for shifting from purely reactive security to a proactive strategy. Analyzing existing attacks has revealed that rapid patching, prioritizing employee security awareness, and adopting continuous threat-hunting tactics are essential. To help you achieve layered, resilient security, the next chapter will focus on the practical deployment of the technologies and approaches mentioned here, setting you up to not just fight today's attackers but to also be better prepared for those of tomorrow.

In the next chapter, we will continue our journey of building a proactive defense program by exploring specific technologies, controls, and best practices for each layer of your defense model, turning the concepts discussed here into a well-structured action plan.

Key takeaways

- Defense is very similar to attack; both are very iterative, and the only difference is the shift in perspectives.

- Network exploits are among the most prevalent attack types in the wild. Attackers from script kiddies to the most sophisticated APTs often exfiltrate systems by sending remote exploits.

- Software injection vulnerabilities have been on the OWASP Top 10 list since its inception [11]. While injection is a very broad vulnerability class, defenses often revolve around special treatment of user inputs.

- Amid information on millions of vulnerabilities sent to organizations, it is challenging to prioritize the right risks. This is a continuous battle and security teams need to make smart decisions on effectively prioritizing risks and mitigating the most critical ones.

- Social engineering, although undermined, is a fundamental component of understanding security. Reading more about this topic develops the deceptive mindset that is often crucial to defending systems.

- Relying on obscurity as a security measure typically proves ineffective in real-world scenarios. When dealing with organizations of 100 or more individuals, determining what remains unknown can present significant challenges.

- SBOM might not be the most effective security artifact as it is designed today, but it pushes us in the right direction toward knowing the dependency chain and being aware of supply chain risks.

- Observing attack attempts on your systems and studying key indicators such as sophistication, persistence, and targets reveals underlying attacker profiles.

Congratulations on getting this far! Pat yourself on the back. Good job!

Further reading

To learn more about the topics that were covered in this chapter, take a look at the following resources:

- [1] Understanding denial of service attacks: `https://www.cisa.gov/news-events/news/understanding-denial-service-attacks`

- [2] Log4j vulnerabilities: `https://logging.apache.org/log4j/2.x/security.html`

- [3] CVE-2021-44228: `https://nvd.nist.gov/vuln/detail/CVE-2021-44228`

- [4] Equifax data breach: `https://www.csoonline.com/article/567833/equifax-data-breach-faq-what-happened-who-was-affected-what-was-the-impact.html`

- [5] CVE-2017-5638: `https://nvd.nist.gov/vuln/detail/cve-2017-5638`

- [6] Pretexting: `https://en.wikipedia.org/wiki/Pretexting`

- [7] Twitter hack update: `https://blog.twitter.com/en_us/topics/company/2020/an-update-on-our-security-incident`

- [8] DEFCON Social Engineering Community: `https://www.se.community/`

- [9] Coverup of Uber data breach: `https://www.justice.gov/usao-ndca/pr/former-chief-security-officer-uber-convicted-federal-charges-covering-data-breach`

- [10] SolarWinds attack: `https://www.csoonline.com/article/570191/solarwinds-supply-chain-attack-explained-why-organizations-were-not-prepared.html`

- [11] OWASP Top 10 injection: `https://owasp.org/www-project-top-ten/2017/A1_2017-Injection.html`

7

Building a Proactive Layered Defense Strategy

In the final chapter of *Part 2*, we'll bring together the insights gathered from previous sections to explore the key elements of a proactive **Defense in Depth (DiD)** security plan. Our aim is to integrate the various aspects of attacker perspectives and create a security framework that effectively counters even the most determined and sophisticated threat actors.

To recap, in *Chapter 3*, *Building a Framework for Layered Security*, we equipped ourselves with a systematic approach to building a robust defense program with organizational risk as our guiding principle. Then, we started *Part 2* by shifting our perspective from a defensive mindset to more of an attacker mindset. We extensively researched the attacker mindset in *Chapter 4* by gaining invaluable insights into their tactics and motivations, followed by our exploration of using an adversarial lens to uncover weaknesses in our systems in *Chapter 5*. Later in *Chapter 6*, we took a 180° turn to observe attacks in real time and link them back to attacker profiles. With all those learnings with us, we are now going to build a proactive security strategy rooted in DiD principles.

First, we'll delve deeply into the fundamental security concept of zero trust. The aim is to lay a strong foundation for constructing a DiD strategy by breaking down the challenge into smaller, manageable components. Next, we'll explore the specific controls constituting the security strategy, examining a range of security measures appropriate for relevant attacker profiles. Additionally, we'll explore the role of automation in streamlining routine security tasks to facilitate ongoing enhancements. Finally, we'll revisit the inherent challenge that complicates security efforts: the inevitability of change.

In this chapter, we're going to cover the following main topics:

- Principle of zero trust
- Selecting the right security controls
- Utilizing SOAR
- Staying on top of changing threats

Let's get started!

Principle of zero trust

Security isn't about perfect prevention, but about making life so difficult for attackers that they give up and target someone easier. Zero trust accomplishes this. By removing implicit trust and assuming compromise is always possible, it dramatically shrinks the attack surface and the impact of a successful breach. Through granular access controls, micro-segmentation, and relentless authentication, zero trust significantly reduces the attack surface and limits the potential devastation a successful breach can cause.

Let's contextualize this concept within another paradigm from computer science known as **divide and conquer**. This algorithmic approach involves continually breaking down a larger problem into smaller, more manageable components until they are easily solvable. Now, envision your organization and its intricate systems working together to deliver value to customers. Protecting these systems presents a significant challenge, partly due to their sheer size.

Traditionally, organizations relied on robust perimeter security solutions to address this challenge. They established façade clients on the public internet, implemented DMZs to route requests appropriately, and once inside the perimeter, allowed data to flow freely between internal systems. However, within this interconnected web of systems, any compromise could result in widespread data exfiltration.

Zero trust aligns with "divide and conquer" by creating numerous smaller security zones, often down to the application level. Just as smaller code chunks are easier to write and debug, each zone has granular access controls, reducing the impact of an attack staying localized. It's no longer one giant problem to defend, but many more manageable perimeter checkpoints.

Core principles of zero trust

Zero trust rejects the traditional assumption that anything behind the perimeter is inherently safe. Instead, it operates on these central principles:

- **Never trust, always verify**: Every user, device, and network connection, regardless of origin (within or outside the perimeter), must be continuously authenticated and authorized. Trust is dynamic, continuously reassessed, and not granted implicitly.

- **Micro-segmentation**: Instead of a single "inside" network, zero trust creates granular zones based on applications, workloads, data sensitivity, and so on. Access between these segments is tightly controlled. This compartmentalization significantly reduces the impact of a breach by containing attacker movement.

- **Principle of least privilege**: Users and applications are given only the bare minimum permissions and access required to perform their intended functions. Any privilege escalation creates red flags for early anomaly detection during attacks that try to obtain greater access.

- **Context-based authorization**: Access decisions are made based on dynamic factors beyond simple identity alone. These can include the following:

  - Time of access attempt

  - Location

  - Device health and security posture

  - Behavior or risk metrics based on threat intelligence

With these insights in mind, let's evaluate the outcomes we expect from a zero trust security model.

Key outcomes

Don't simply buy into the hype surrounding zero trust. It's crucial to grasp its actual benefits. Essentially, by eliminating implicit trust between systems, zero trust enforces authentication across all layers. However, it's vital to note that if authentication or authorization is mandated at every interface and a shared static secret is used, the efficacy of zero trust vanishes. Understanding the added overhead is essential to prevent the creation of workarounds that compromise the security model. Let's explore some key outcomes we expect from a zero trust framework:

- **Reduced attack surface**: Limiting privileges and network segments makes lateral movement more difficult, minimizing the potential for system-wide compromise.

- **Increased visibility and monitoring**: A strong identity foundation and fine-grained access controls allow for better anomaly detection, even of threats operating from within.

- **Improved adaptability**: A zero trust approach provides better resiliency by facilitating changes to security protocols in response to new threats and business needs – without sacrificing overarching security principles.

Zero trust transforms security from a static model into a continuously adaptive process. Instead of blindly trusting systems or people once they are granted entry, every action is monitored and assessed based on its need. This fosters a mindset of proactive investigation and containment of threats, making it dramatically harder for attacks to be successful at scale.

We've discussed the fundamental principles of zero trust. Now, let's delve into the practical implications of these principles within real-world systems.

Impact of "never trust, always verify"

What does it really mean to authenticate at every step? Let's look at some benefits of this:

- **Internal breaches contained**: Continuous authentication and authorization limit the ability of insider threats (compromised accounts, malicious actors), minimizing damage if a single point of entry is breached.

- **Attacks slowed down**: Attackers can no longer assume "getting in" allows them unrestricted movement. Having to constantly renew/escalate permissions increases the risk of detection.

- **Defense focus shifts**: Less effort needs to be spent on the perimeter once it becomes only one of many checkpoints, freeing resources for proactive defense.

Impact of micro-segmentation

The benefits of dividing your architecture into smaller parts can be many, but let's look at some obvious ones:

- **Limiting lateral movement**: Even sophisticated attackers who achieve an initial foothold struggle to "pivot" between systems when fine-grained permissions are in place. This buys valuable time to respond.

- **Reducing systemic risk**: Compromising a low-value, micro-segmented system no longer automatically grants the keys to everything else. Each breach is essentially isolated to its own zone.

- **Enabling visibility**: Network flows between segments are far easier to monitor than within the previously "flat" perimeter, revealing attacker tactics early.

Impact of least privilege

The principle of least privilege dictates that users, accounts, and processes should only have the absolute minimum privileges and access necessary to perform their intended functions.

- **Reducing exposure**: Even compromised accounts remain tightly constrained, restricting what malicious code can achieve if an attacker leverages a valid user to launch a broader attack.

- **Limiting "blast radius"**: A breach is less likely to immediately lead to wide-scale data exfiltration due to the strict, localized application permissions.

- **Improving logging and detection**: Anomalous actions and privilege escalation attempts stand out when most activity operates within clear baselines of necessity.

Upon closer examination, you'll notice a recurring theme among these principles: the emphasis on minimizing and segmenting the attack surface. Regardless of the complexity of your underlying systems, a zero trust approach entails decoupling security properties to some degree. This forms the basis for establishing a DiD security strategy.

Now that we have established a common understanding of the core principles, let's explore some practical implementations to deepen our understanding.

Practical implementation of zero trust

Translating the concepts of zero trust into operational action involves technology, process changes, and a cultural shift to assume compromise is possible. When implementing a zero trust model for your organization, begin by reevaluating your network architecture and access controls. This involves moving away from traditional, perimeter-based security models and adopting a more granular approach to access control. For example, instead of relying solely on network segmentation, organizations implement micro-segmentation to create smaller, isolated network segments based on workload or application sensitivity. Some key strategies to achieve this are covered in the following sections.

Identity and access management (IAM)

IAM can mean many things. In the cloud-native world, it can mean a product, but in general, IAM is all about authentication and authorization, as we talked about in *Chapter 1*. The motto of zero trust is to authorize every request and there's no implicit trust built into the system. Here's what you can do to get started:

- **Robust authentication and authorization**: Enforce **multi-factor authentication** (MFA) and adaptive authorization rules based on context (device trust, threat level, location). This strengthens front-line access and provides dynamic access control.

  Example: Low-risk actions within business hours trigger basic authentication, and unusual accesses (from new devices/locations) prompt escalated MFA.

- **Privileged access management** (**PAM**): Tightly control administrative accounts, requiring justification for usage, session monitoring, and time-bound access, limiting their exploitation potential.

  Example: Critical production components with customer PII will have stricter access control. Persistent access to these environments might not make sense. Time-bound access via break-glass mechanisms is widely used in the industry.

Access on demand is a relatively new idea that has been gaining popularity recently. As we want to move away from having persistent access across systems, this idea can be your trust anchor. In an ideal production deployment, you would not want to have anyone with persistent access, but could grant access for a shorter duration of time when someone needs to get in to debug or fix an issue. In reality, for disaster recovery purposes, you likely will end up with a few admins, but consider them as your point of failure when designing security. DiD will come in handy here.

Next, we will look at dividing our security problem with network ACLs.

Network access controls (NACs) and micro-segmentation

Gone are the days when we used to segregate our application environments into two or three network segments (external, internal + untrusted, and internal + trusted). As security becomes a priority, we start off by making logical divisions at the company level. But that's not enough; we are at the technological juncture where we need to readjust our methodologies once more. Let's explore some options:

- **Network segmentation**: Subtly divide systems based on sensitivity and business need, not just general "inside" versus "outside." Use firewalled DMZs, VLANs, or advanced **software-defined networking** (**SDN**) architectures.

  Example: Isolate the HR systems, requiring MFA even from the intranet, as it holds private employee data.

- **Zero trust NACs**: Enforce granular access policies beyond simple IP, incorporating user identity, device state, and so on. Contextual awareness blocks compromised or insecure devices from connecting.

As companies are moving toward cloud-native, it is important to keep a **default deny** policy in mind for your networks. You may be surprised to hear how many companies are "accidentally" making their S3 buckets public. [1]

Device trust and endpoint security

Like security, trust is very similar to a chain, and with every chain, you need an anchor. In the world of security, this is commonly referred to as the **root of trust**. Establishing the trust anchor is a very hard problem to solve. Let's take an example to solidify our understanding of the problem here, and think of **public key infrastructure** (**PKI**) systems that ensure trust across the whole web. When we connect to an unknown remote server, our browser is doing the leg work to verify the authenticity of the server by validating a certificate it presents (subtly assumed using HTTPS). But the question is: why did our browser trust the certificate? It was likely because it was signed by an entity (certificate authority) that we trust, or the trust is somewhere along the chain.

In the world of computes and organizations, often, the root of trust is built from the hardware that runs our applications. Ensuring endpoints that perform privileged actions are trustworthy is key to any defense strategy. Here, we will explore a couple of ways to get started:

- **Device posture assessments**: Verify patch levels, configuration checks, and security agent presence before allowing connection. This creates a more hygienic network overall, not just zero-trust-focused.

- **Endpoint detection and response** (**EDR**): This actively monitors devices, not just scans. It utilizes behavioral analysis and threat intelligence feeds to alert on suspicious activity in real time, ensuring internal breaches stay small.

With the rise of remote work, **bring your own device (BYOD)** policies, including techniques such as Apple's **mobile device management (MDM)** [2], are being widely adopted.

Security orchestration, automation, and response (SOAR)

SOAR unifies the various security tools an organization uses (firewalls, endpoint protection, threat intelligence feeds, etc.). It helps streamline and improve incident response activities. SOAR is often used interchangeably with automation, but it's much more than that. We will first break down the term and explore each part:

- **Orchestration**: SOAR coordinates actions across different tools. Instead of analysts juggling multiple systems, a single alert in SOAR can trigger a coordinated response chain across tools you already own.

- **Automation**: SOAR can automate repetitive security tasks such as the following:

  - **Alert triage**: Prioritizing alerts based on severity and potential risk.

  - **Threat intelligence lookups**: Automatically enriching alerts with external threat data.

  - **Initial containment**: Blocking malicious IPs and quarantining infected machines.

- **Response**: SOAR provides dashboards and playbooks (guided procedures) to facilitate incident investigation and response. Analysts gain better visibility and a central command point.

The benefits of adopting SOAR in your security program can be multi-fold. By eliminating the manual time spent on repetitive tasks and accelerating actions to isolate and neutralize threats, you can reduce response time, making your IR more efficient. Keep in mind that someone is likely spending their time doing these tasks and often, a SOAR solution frees up defenders from being mired in routine alerts, allowing them to focus on higher-order investigation and threat hunting, saving the business money. Another subtle benefit of SOAR is consistency. Playbooks ensure a consistent incident response, even across shifts or different personnel levels. Let's quickly go over how to get started:

- **Automated responses**: SOAR integrates threat feeds, access logs, and so on to automate initial breach responses or alert relevant teams quickly, reducing manual effort and improving reaction time.

- **Enhanced visibility**: SOAR connects diverse data sources across your security stack, enabling threat hunting and anomaly detection that a non-zero-trust model is less equipped for.

Remember, SOAR is an iterative process as a defense program matures. As you identify your most time-consuming, repetitive security tasks, SOAR enables you to streamline them. This not only improves efficiency but also frees up your team's bandwidth to focus on proactive strategies and address sophisticated threats that automation can't fully contend with. Later in this chapter, we will dive deep into this topic with real-world examples.

We've looked at some key strategies for implementing zero trust; next, we will take a step back to understand how zero trust fits into the DiD model. Zero trust fundamentally restructures how we understand the web of systems. By breaking down the monolithic "inside" perimeter into smaller, micro-segmented zones, it essentially transforms each micro-segment into its own miniature, fortified system. With access now tightly controlled on every boundary, each application or workload becomes a distinct defensive layer in the grand scheme.

The beauty of this model is that DiD principles now operate with far greater granularity. Within each smaller segment, traditional security controls become even more effective. **Network intrusion detection systems (NIDSs)** are used to monitor traffic. Here, we can focus on a manageable stream of traffic instead of the once-overwhelming torrent within the wider network. Host-based firewalls, application-level whitelisting, and rigorous privilege control now have clearly defined domains to govern – making anomalies easier to spot.

Furthermore, zero trust's continuous authentication philosophy is perfect for layered defense. Every transition between segments can become a checkpoint for validation, creating additional hurdles for an attacker at each step. Even after establishing a foothold in one segment, lateral movement faces new authentications and permission checks, effectively turning "layers" into numerous smaller battlefields where a breach can be swiftly contained.

Remember, zero trust doesn't erase the need for traditional DiD but enhances it from the inside out. With each microsystem having its own layers of protection, even a successful breach becomes less catastrophic and more of a focused intrusion that a response team can effectively counter. This subdivision essentially creates layers of security control for attackers to penetrate, thus forming the basis of DiD.

Next, we will walk through a real-world zero trust framework that was developed at Google after Operation Aurora [3].

BeyondCorp – A real-world case study

Operation Aurora was a multi-faceted, targeted campaign launched in mid-2009, primarily against Google but also compromising over 30 other high-profile technology and defense sector companies. Investigators suspected the attacks were state-sponsored and originated from China [4].

What happened?

The attackers' initial access came via spear-phishing emails. This led to targeted malware exploiting a zero-day vulnerability in Internet Explorer. Attackers were able to maintain persistence inside Google's internal network for an extended period, exfiltrating intellectual property and compromising the accounts of human rights activists. Following Google's disclosure of the attacks in January 2010, a wave of similar breaches impacted other corporations.

Google's decision in January 2010 to publicly disclose the Aurora attacks was pivotal. It revealed the sophistication of the actors and alerted other organizations to similar ongoing intrusion attempts. Subsequent investigations confirmed Aurora had breached over 30 additional companies, primarily within the technology, aerospace, and defense sectors. Adobe, Juniper, Rackspace, and many others experienced similar attacks.

Aurora demonstrated the dangers of zero-day attacks and the need for vigilance regarding supply chain security. Companies had to rapidly change how they assessed risks from unknown vulnerabilities, and not just patch known ones.

While attacks such as these happen continually, Aurora remains a seminal cyberattack case study because it forced organizations across the globe to rethink their defense strategies. The attack underscored that even leading tech companies were vulnerable, highlighting advanced, persistent threats lurking behind seemingly innocuous phishing emails. Further, it put zero-day exploits in the spotlight and demonstrated the necessity to actively hunt for threats inside the network instead of purely focusing on perimeter defense.

Lessons learned

It's worth noting that lessons learned from high-profile breaches such as Aurora are ongoing. However, this landmark event undeniably shifted the industry's focus to anticipating persistent threats and investing in proactive detection technologies. Here, we will highlight some shifts in the security mindset:

- **Zero-days pose extreme risks**: The attackers' use of a previously unknown Internet Explorer vulnerability highlighted that traditional, signature-based defense could be rendered ineffective. Vigilant patch management remains vital, but cannot protect against these advanced attacks alone.

- **Insider threat isn't just malice**: Even well-intentioned employees can be attack vectors due to targeted social engineering. Beyond simple password policies, security training must emphasize attack techniques such as spear phishing and the risks of unknown malicious links.

- **Persistence is key to stealth**: Attackers were focused on low-and-slow data extraction in Aurora, not quick disruptions. Organizations needed to re-evaluate detection to spot small anomalies in normal traffic patterns, assuming breaches would occur on some level.

- **Defense must be proactive**: Waiting for an attack to reveal itself proved insufficient. The emphasis grew on proactive threat hunting for signs of intrusion within company systems, utilizing behavioral analysis and threat intelligence sharing in a way that wasn't done pre-Aurora.

- **Supply chain is an exploit vector**: Although difficult to confirm definitively, the potential compromise of Adobe's code-signing tool emphasized that attacks on software creators could be leveraged to spread to their customers. Rigorous review processes for third-party code in the supply chain became paramount.

Amid the challenges presented by the hypothetical worst-case scenario, it's essential to acknowledge the positive aspects of the broader cyber landscape. Let's take a moment to highlight the successes and achievements that have contributed positively to our overall cybersecurity posture:

- Google pioneered the sharing of details of these events publicly and collaboration with authorities. This encouraged more information exchange between impacted companies and law enforcement, though such communication often remains slow and complex.

- Aurora accelerated development in technologies focusing on network behavior analysis, EDR solutions, and automated threat intelligence sharing. Organizations could no longer rely solely on perimeter defenses as attackers were actively operating inside networks.

As **advanced persistent threats** (**APTs**), nation-sponsored cyberattacks emerged, companies started collaborating more closely in defending against the most persistent attackers. Now, let's look at how it relates to zero trust.

Inception of BeyondCorp

Let's explore Google's BeyondCorp initiative, a bold re-imagining of corporate network security born in response to the lessons learned from Operation Aurora and similar breaches.

Attacks such as Aurora showed that the "inside network = trusted" concept was outdated. Attackers could compromise even secured intranets if successful in the initial user compromise (e.g., spear-phishing attacks). Traditional access models with static firewalls became inadequate against attackers with internal footholds. Granting trust based on location is an ineffective defense if access credentials or devices within the perimeter are stolen.

Keeping these changes in mind, BeyondCorp started off as a fundamental shift in how security architecture was re-imagined. The BeyondCorp approach included the following elements:

- **Zero trust inside**: BeyondCorp extends zero trust principles ("never trust, always verify") to every device and user attempting access, regardless of whether they are already "inside" the network.

- **Device- and user-centric**: Authentication and authorization rely heavily on granular device identity (patch status, running security agents) and user context (role, location, risk metrics).

- **Shift to access proxy**: Network access is mediated through the access proxy. This enforces continuous policy evaluation based on dynamic factors that may indicate compromise beyond mere authentication credentials.

- **Unified threat/data protection**: Access to corporate applications is intertwined with Google's globally deployed protection against malware, DDoS attacks, and data exfiltration attempts.

The shift here was toward "managed" devices, which are company devices with identities embedded within them. Communication with any internal resource would go through an access proxy, which authenticates the request based on the device it originates from. As a result, the network lost its relevance.

Here's an overview of what BeyondCorp looked like when Google introduced it:

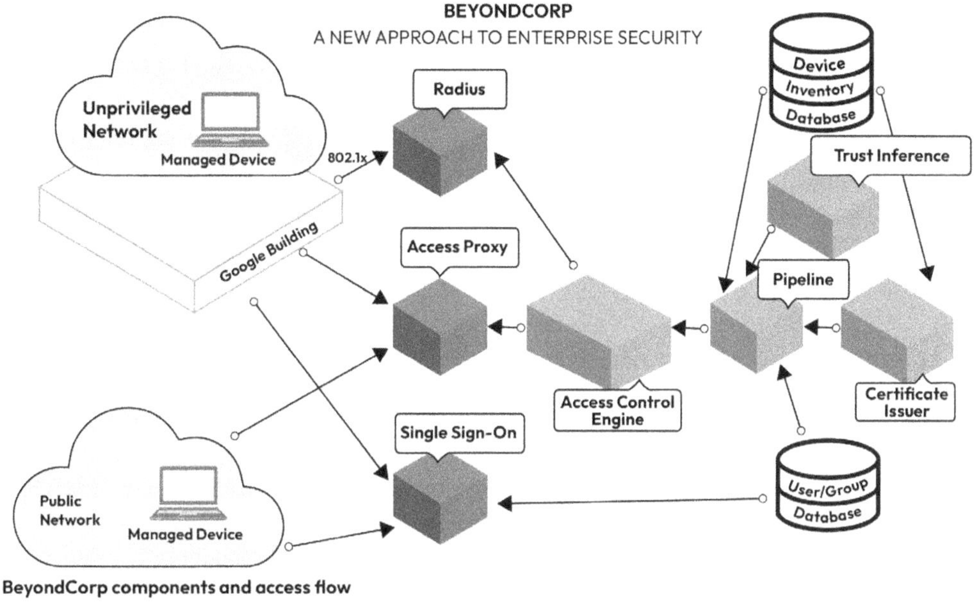

Figure 7.1 – BeyondCorp architecture (source: https://www.beyondcorp.com/)

As a result, traditional VPNs become redundant when "being inside" no longer equals inherent trust. This simplifies remote work while improving security. It also means more control over precisely what each user or device can access based on dynamically updated metrics. While initially pioneered by Google, zero trust has grown industry-wide. BeyondCorp was a key driving force behind Google's internal migration toward this model, showcasing how such radical shifts are achievable, even for companies as large as Google itself.

Designing attacker-informed defense

Our journey so far has led us into the attacker's playbook. Whether dealing with opportunistic script kiddies exploiting a widespread software flaw (such as Log4j) or a targeted state-sponsored group carefully infiltrating a defense contractor, we've learned that analyzing adversaries isn't about getting inside their heads – it's about exposing their tactics. We learned the hard way that assuming that "good-enough" security equates to real protection is a disastrous fallacy.

The zero trust framework we've explored marked a critical shift toward continuous evaluation. Attackers don't take breaks, and neither can our defenses. However, zero trust alone isn't enough. Building robust micro-segments and enforcing strict authentication is no replacement for understanding how a determined attacker exploits the cracks left behind, even in well-structured networks.

Now, it's time to turn the tables. Imagine attackers scrutinizing your organization as deeply as we've investigated their methods. This section focuses on leveraging those KPIs – the sophistication, targets, and persistence we discussed in *Chapter 6* – to gain true defensive power through informed control choices and prioritization. It's not just about blindly implementing the latest security tool, but about intelligently deciding which controls will have the most disruptive impact on attacker success.

DiD isn't just layers of different technologies. With proper awareness of attacker mentality, even seemingly small decisions, such as patch cadence and configuration choices, become opportunities to disrupt malicious processes that attackers rely on to be effective. In doing so, we go beyond a checklist mindset toward resilience – making attackers work far harder than the potential payoff ever warrants. Let's dive in!

Zero trust – Good start, not foolproof

Zero trust is a good start, but it is important to remember the term "zero trust" lost some of its meaning with the latest hype around it. In the earlier section and throughout this book, we have highlighted how important it is as a defender to avoid buying into the hype without understanding the fundamental idea. It is very important to first understand the problem before inserting a solution. The effectiveness of a solution often lies within understanding the initial problem and the limitations of it.

Hypothetically, let's imagine a company celebrating a zero trust architecture, yet not enforcing code reviews on the application code that will eventually make its way into production. This is a very common scenario across the industry, and it is very important to understand that a "zero trust model" is no magic wand that will make your security problems go away.

Attacker's eye view

Adopting zero trust gets you ahead, but doesn't mean attackers cease trying. Zero trust undeniably strengthens security posture, but its effectiveness lies in meticulous implementation. Like in any complex architecture, even well-intentioned efforts can leave gaps that astute attackers will find and exploit. It's important to remember that while a zero trust mindset drastically reduces your attack surface, threats continue to evolve, requiring regular refinement.

Here are a few common missteps and their potential consequences:

- **Inconsistent policy enforcement**: Dividing the network into small zones is powerful, but only if granular isolation and controls are enforced. Leaving "bridges" between segments due to overly complex policy structures, or failure to monitor traffic traversing zones, allows attackers to capitalize on a foothold by moving laterally with deceptive ease.

  For example, development environments containing test data often have relaxed access to mimic production to ease developer tasks. Yet, inadequate segmentation from the core network makes that test data, if compromised, a stepping stone for privilege escalation across zones that an actual breach would struggle with.

- **Legacy system blind spots**: While prioritizing new applications for segmentation, neglecting old servers and infrastructure elements creates gaping holes. These often have insecure configurations and run unpatched software, but have broad network access due to "always being there." They offer easy initial entry, undermining the whole model. Legacy systems are often hidden in the inner layers of a complex system, just like layers of onion.

Figure 7.2 – Comic representation of a legacy blind spot

A good example here could be the SolarWinds attack, in which the compromise partially arose from attackers breaching management software for servers, then pivoting through a vast, poorly segmented network of SolarWinds clients.

- **Implicit trust within zones**: Zero trust should extend deep inside systems. Yet, granular authorization is harder on applications themselves. Failing to apply "least privilege" to applications within microsegments allows attackers to escalate from a simple app compromise to wider control, since the app has overt permissions within the system where it resides.

  For example, we could design a system where a user request is authenticated at the frontend and once authenticated, all backend components are open. This is where a lot of confused deputy problems arise. One compromise can lead to the entire system being subverted.

These mistakes exemplify why implementing zero trust needs to be paired with the mindset we've built. The key is continuous re-evaluation of policy, vulnerability scanning even within trusted segments, and vigilance toward the assets that attackers would view as the weak links in your new "hardened" infrastructure.

Controls with attacker disruption in mind

Revisiting our earlier dissection of attacks such as network penetration and web application exploits, we now wear the attacker's "hat." After landing that initial exploit (perhaps by injecting malicious code into an unsanitized user input field), the question now becomes: how do I turn this access into something valuable? Our controls are designed to make their next steps painful, uncertain, and time-consuming. Let's explore some choices here.

Control choices through impact

Whenever we come across such questions, we need to ask what's most disruptive to attackers. Let's walk through some examples to solidify our understanding:

- **Input sanitization**: It may seem obvious, but its true importance becomes magnified. Proper input sanitization is not about making an attack outright impossible but about limiting its blast radius. When **cross-site scripting** (**XSS**) can't execute injected JavaScript, attackers can't easily jump beyond, say, defacing a blog comment, and need to find a new method of escalating privilege. If zero trust segmentation is solid, they're back to square one in trying to escape that initial, low-privilege sandbox.

- **Application reviews**: These become a hunt for excess permissions or internal logic flaws. It's a recognition that even secure code is rarely static. If a compromised account within a given segment has wide-ranging abilities, attackers may never need to get out of their micro-network zone as there's plenty to exfiltrate or disrupt before defenders notice. A code review with "disruption potential" in mind ensures that even an otherwise secure web application isn't an attacker's free buffet after finding one way in.

  Application reviews are not just for bug fixes, but to spot privilege creep within software, limiting what an attacker can exploit within even compromised code. Chromium regularly prunes code ownership privileges from inactive owners; this is an attempt to ensure that likely inactive accounts don't get used for compromising the source code [6].

- **Network traffic analysis**: Zero trust creates the ideal hunting ground, not an obstacle. Anomalous behavior sticks out because every microsegment should have well-defined communication patterns. When an attacker, even if they are successful initially, needs to do network reconnaissance to figure out where to go next, unusual traffic spikes and port scans expose them. Visibility within our internal systems isn't about being perfect upfront but about catching attackers making noise as they stumble around with limited knowledge.

The examples in this section were carefully crafted to not provide a coherent list of things to do. The idea is to pay attention to the reasons behind selecting one control over another, and to look again at the exploits, but do so from the attacker's viewpoint: "If I used XSS here, what would prevent me next?" This will form the base for creating a resilient DiD security model.

Defense in depth, evolved from the inside

Throughout this chapter, our discussions consistently circle back to the concept of security controls at various layers. Whether we're exploring the zero trust model or considering an attacker's viewpoint on segmented systems, each concept underscores the importance of defending individual components of an application in the event of a breach. However, our emphasis lies in understanding the aftermath. The primary objective for defenders is to minimize the impact of attacks, recognizing that they are inevitable.

The layered mentality, leveled up

DiD has always been about making attackers fight uphill – each layer they must penetrate should exhaust more of their resources. Zero trust supercharges this. The strong "inner walls" of microsegments force attackers to solve new, smaller puzzles as they progress. Even one misconfiguration won't automatically spell compromise due to those compartmentalized barriers. Each layer now requires new tools and new exploits, increasing their "cost."

As modern defensive systems adopt these principles by design, a growing sentiment across a part of the security community is that the common phrase that "the attacker only needs to be right once" is fading. While there's some merit to that point of view, we should remember that an attacker's toolkit is also becoming ever-powerful. There are so many unknowns we deal with when it comes to security, so it is very important to remember that an attacker still needs to find the weakest link. Now, if you compare the weakest link from the modern day to that of 30 years ago, of course it will be different.

With the heavy adoption of **machine learning** (**ML**) and especially GenAI, attackers' weapons are becoming not only modern but also smart. To stay ahead of the curve, we need to start thinking about reproducible, segmented, and layered defense mechanisms. It's time to level up!

Proactivity versus checklist

Compliance mandates are a base, but we're focusing on what demoralizes attackers. Prioritizing a control for its disruptive potential (such as meticulous input sanitization) can force an attacker who specializes in one exploit type back to their research phase because what previously worked is now ineffective. This dynamic evolution, informed by the attacker's viewpoint, ensures your defense posture keeps adapting even as threats do, always maintaining an element of surprise for those targeting you.

Attack breakdown and defense takeaways

Let's map a particular attack chain on a well-defended network to strengthen our DiD thought process. We'll use the 2020 Twitter breach, as its stages became clear through subsequent security probes.

Phase 1 – Social engineering – Spear-phishing

Attackers targeted Twitter employees who had access to internal support tools via elaborate, phone-based impersonation schemes (aimed to obtain credentials/access tokens). Even with training, it happens; for example, the user clicks a malicious link in an email.

Things that could have gone better are as follows:

- Even high-value tools deserve isolation. Strict segmentation/controls at every authentication point prevent one compromised employee account from granting unfettered access.

- Security training needs tailoring (this was *not* simple password trickery) depending on the threats you face. Attacks on user empathy and urgency creation must be emphasized.

- A strong outer perimeter remains crucial. While zero trust helps later, stopping malicious access initially remains ideal.

Phase 2 – Internal tool manipulation

Once the attacker gained the initial foothold in the restricted network, with legitimate tools access, they easily manipulated user data, changed profile info, and likely sought high-profile targets based on internal info that was visible with these privileges.

Things that could have gone better are as follows:

- Internal tools often have unnecessarily high permission. Regular auditing of what is truly needed is key. Unauthorized profile actions should send immediate red flags in this scenario.

- Log *everything* that those powerful tools do. If zero trust fails upfront, having records enables a forensic response, limiting attackers' freedom to operate unseen.

Phase 3 – Tweet propagation

Attackers used compromised high-profile accounts (Musk, Gates, etc.) for a Bitcoin scam. This phase became visible, but prior access was likely used for reconnaissance to select "juicy" targets.

Things that could have gone better are as follows:

- Even legitimate admins sometimes do unusual things for support. An easy flag that requires human verification when specific account tiers are touched can delay abuse. Adding a second human for approval often attached to a shareable ticket can add layers of security to the process.

- Out-of-character tweets should have low tolerance. AI alone misses context for satire, but this context exists within the company – employees are likely to spot a rogue Musk tweet faster than signature-based scanning.

The Twitter hack had many complexities. Some defense points here overlap – zero trust alone could not have stopped all phases, but it would have been the start of a DiD strategy from the inside.

In an earlier section, we briefly touched on SOAR. Next, we will explore how we can utilize SOAR to advance our defense to the next level.

Utilizing SOAR

While zero trust principles provide a rigorous framework for authentication and access control, and DiD offers a layered security posture, a SOAR platform serves as the operational catalyst that brings these concepts to life. SOAR solutions actively translate those strategies into dynamic workflows and automated responses. By empowering security teams to detect, investigate, and counter threats with unrivaled speed and consistency, SOAR delivers the tactical edge to enhance defense resilience.

In this section, we'll explore how SOAR can be deployed to tackle some of the most pressing challenges facing security operations. Through real-world use cases, we'll illustrate how SOAR automates crucial processes within a zero trust environment, streamlines responses, and minimizes the operational burden on analysts. You'll gain insights into how SOAR platforms bolster your existing DiD, proactively identify threats, and provide enhanced agility and efficiency across your cybersecurity operations.

Real-world SOAR defense use cases

Let's turn our attention to real-world use cases to understand the challenge and the value that SOAR brings to a defense program. We will explore three use cases that are very commonly used in the industry and walk through the benefits that such a platform will bring.

Use case 1 – Proactive threat hunting

Manually sifting through massive volumes of security data to identify subtle threat indicators is time-consuming, error-prone, and often, reactive. Threat hunters need proactive tools to stay ahead of adversaries.

Now, let's see how SOAR can help here with a sample workflow:

1. **Data collection**: SOAR continuously ingests data from SIEM, EDR, threat intelligence feeds, and external sources.

2. **Threat indicator enrichment**: It enriches identified **indicators of compromise** (**IoCs**) such as file hashes, IP addresses, and domains with contextual information (reputation, malware associations, etc.).

3. **Hypotheses generation**: It then applies ML or pre-built playbooks to correlate disparate data, uncover patterns, and generate potential threat hypotheses.

4. **Prioritization and alerting**: SOAR prioritizes hypotheses based on risk scores and automatically alerts analysts to high-probability threats.

5. **Guided investigation**: SOAR provides a centralized workspace with enriched data and relevant tools, guiding analysts through efficient investigations.

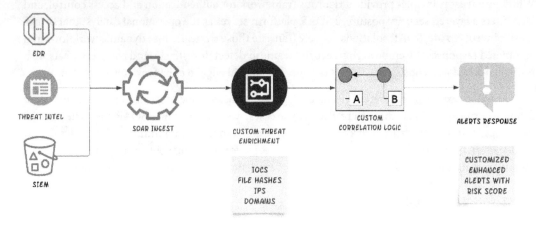

Figure 7.3 – Example of a threat-hunting SOAR workflow

With this very simple workflow, SOAR streamlines the threat-hunting process, uncovering potentially dangerous activity that might have otherwise gone undetected. This significantly reduces dwell time and the potential impact of breaches.

Use case 2 – Phishing attack response

Phishing attacks bypass perimeter defenses with increasing sophistication. Rapid response is essential to minimize damage and user compromise. SOAR is increasingly being utilized to respond to spam emails found in corporate mailboxes.

Here's a sample workflow for our study:

1. **Alert trigger**: SOAR integrates with email gateways and user reporting tools. A reported phishing email triggers an automated response.

2. **Incident creation**: It creates an incident ticket with all relevant details (email headers, links, attachments, etc.).

3. **Detonation and analysis**: SOAR can detonate attachments in a sandbox and analyze URLs for maliciousness. This can then be fed into a list of known malicious links.

4. **Quarantine and search**: Then, it quarantines suspected malicious emails and searches the entire environment for identical instances.

5. **Remediation and user education**: SOAR facilitates the deletion of phishing emails across inboxes and delivers targeted user education to those who have interacted with the email.

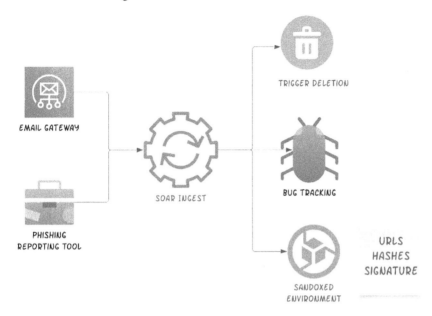

EMAIL GATEWAY

PHISHING
REPORTING TOOL

SOAR INGEST

TRIGGER DELETION

BUG TRACKING

SANDOXED
ENVIRONMENT

URLS
HASHES
SIGNATURE

Figure 7.4 – Example of a phishing-attack SOAR workflow

SOAR dramatically reduces the time needed to contain and remediate phishing attacks, preventing successful phishing campaigns from inflicting lasting damage.

Use case 3 – Insider threat mitigation

Before closing out our exploration, let's look at one last use case where SOAR can really make a difference. Insider risks are notoriously difficult to detect. Malicious insiders often have legitimate access and know how to disguise their activities. Traditional security tools struggle to recognize complex behavioral patterns that may indicate malice.

We can use SOAR to connect the dots from multiple tools using correlation and develop a workflow as follows:

1. **Baseline user behavior**: SOAR can facilitate gathering data from a range of sources (logon activity, file access, network usage, etc.) to establish normal behavior patterns for each user and role.

2. **Anomaly detection**: SOAR can employ behavioral analytics and ML models to identify deviations from established baselines. Unusual activities, such as accessing sensitive data outside working hours, excessive data downloads, and unauthorized attempts, raise flags.

3. **Risk scoring**: SOAR dynamically assigns risk scores to user actions based on severity, context, and previous anomalous behavior.

4. **Prioritized alerts**: When risk scores exceed defined thresholds, it can generate alerts, providing analysts with early warnings of potential insider threats.

5. **Contextual investigation**: SOAR provides a central workspace for investigations, consolidating relevant data, offering potential correlations, and suggesting the next investigatory steps.

6. **Intervention**: Additionally, with the help of SOAR, we can automatically take protective measures depending on policy, such as disabling accounts, quarantining devices, and triggering more rigorous authentication.

To clarify this workflow, a SOAR platform generally provides the capability to define these workflows that connect multiple security tools to perform actions; the platform does not take those actions. In this example, SOAR enables the early detection of suspicious activity that could signal insider threats. Analysts have the intelligence to act before severe damage occurs, greatly enhancing data security and the overall cybersecurity posture.

Keep in mind that the previous examples are simple illustrations of the hidden power of these platforms. These playbooks can be customized to your organization's specific needs. With the help of ML, we might be able to auto-enhance the effectiveness of these workflows without too many explicit predefined instructions. However, I'd be a little cautious before going there, as it'd be prudent to avoid marketing gimmicks.

Integrating SOAR for enhanced resilience

Beyond standalone use cases, SOAR shines as the orchestrator and connective tissue of a resilient defense architecture. By seamlessly integrating with existing security technologies such as SIEM, EDR, threat intelligence platforms, and vulnerability scanners, it unlocks even greater value. Think of it as the conductor of a cybersecurity symphony, ensuring that each instrument is in tune and playing the correct part for overall harmony.

One hidden benefit of this integration is a feedback loop. As SOAR handles investigations and automates responses, it feeds insights back into other security tools. For instance, by detecting a previously unknown malicious domain during a phishing response, it can proactively update firewall rules or web filtering systems to block future access. This creates a continuously learning and adapting defense infrastructure.

Beyond adaptability, the power of customizations results in exponential impact. SOAR platforms are often built as a combination of smaller logical units with the option to build your own in-house component from the ground up. This flexibility makes SOAR applicable to most existing security programs. As we enable these platforms to automate certain actions, such as adding a new IP rule to our WAF rule automatically as it detects too many failure attempts, we unlock our defense monitoring to not only core business hours but beyond.

SOAR also enables dynamic configuration updates across the stack using custom plugins. When a vulnerability is identified, SOAR can leverage asset management integrations to instantly understand which systems are impacted. It can then trigger automated patching scripts or push updated firewall configurations as an immediate countermeasure. This ability to act on real-time information vastly expedites remediation efforts compared to manual processes.

The agility offered by SOAR is key to resilience. As the threat landscape shifts, playbooks can be quickly modified and new ones created. When combined with a strong, standardized incident response process, SOAR ensures your teams are equipped to respond decisively to even unforeseen attack vectors, minimizing operational surprises and the potential for chaos during cyber incidents.

One thing to keep in mind, though, is that automation isn't the end goal. One major limitation is its reactive nature. To automate a workflow, you need to have responded to a similar threat before. In the following section, we will switch our perspective to understand some of these gaps that will form the basis of what we are going to cover in the next chapter.

Defense as an open loop

The world of cybersecurity offers no guarantees; even the most robust defense strategies aren't impervious. This chapter acknowledges the ever-present "open loop" risks. We know that zero-day vulnerabilities surface, attackers continuously refine their tactics, and no amount of technology can perfectly predict every threat. In this section, we shift our focus from purely reactive tactics to strategies rooted in proactive resilience.

We'll highlight the significance of proactive monitoring, emphasizing the fundamental necessity of deep visibility into network activity and system behaviors. Real-time intelligence feeds combined with ongoing vulnerability assessments give you valuable insights into evolving risks. We'll also explore the value of extracting lessons from emerging threat data and case studies, ensuring your team consistently adapts its knowledge and responses.

This section underscores that defense resilience isn't a single solution or a state of absolute security. Instead, it's a continuous practice rooted in relentless vigilance, proactive risk identification, strategic updates to defensive approaches, and a deep understanding of your organization's unique digital landscape.

No defense is 100% airtight

Zero-day exploits, such as the infamous Log4j vulnerability, serve as stark reminders that no matter how robust your defenses are, the potential for unforeseen weaknesses always exists. They expose the inherent "open loop" nature of the cybersecurity landscape. A zero-day vulnerability lingers as an unknown flaw in widely used software, allowing attackers to slip through before a patch is even developed. Therefore, relying solely on existing protection mechanisms creates a dangerous blind spot.

To combat this ever-present threat, proactive strategies become paramount. This includes subscribing to continuous threat intelligence feeds that alert you to the latest discovered vulnerabilities and attack techniques. Moreover, regular penetration testing and vulnerability scanning play a vital role in actively seeking out potential weaknesses within your own systems. It's a proactive approach to uncovering hidden gaps in your defenses before adversaries can exploit them.

Evolving attacker methodologies

Attackers, unlike traditional armies, do not fight fair. They constantly evolve their tactics to circumvent conventional defenses, seeking unforeseen openings and pushing the boundaries of traditional attack methods. Recent threat data and case studies highlight alarming trends that demand continuous learning and adaptation throughout your cybersecurity teams. The notion that investing in the latest firewalls and intrusion detection systems is sufficient now fades against this adaptable enemy.

Let's consider the evolving landscape of ransomware. Initially, attacks relied on indiscriminate "spray and pray" campaigns. Today, ransomware has gone corporate. Ransomware groups now invest in complex reconnaissance, weaponizing tactics such as APTs as they map out networks and carefully target high-value data. Additionally, social engineering tactics have transformed phishing attempts from poorly written emails into finely tuned messages targeting specific individuals with carefully crafted impersonations and contextually relevant lures.

This ongoing evolution of tactics underscores a fundamental truth: cybersecurity resilience hinges on the knowledge and adaptability of your people as much as it does on your technological defenses. Incident response procedures need to be dynamic, mirroring the agile approaches of adversaries. Analysts must continuously digest fresh threat intelligence, incorporating new tools and methodologies to counter sophisticated threats. Continuous training and cross-team knowledge sharing foster an environment where learning rivals threat development in a perpetual race to anticipate and outmaneuver attackers.

Summary

After exploring the attacker mindset over the last few chapters, here, we wrapped our heads around building a proactive defense strategy. We started off with an in-depth exploration of zero trust security principles, followed by its key benefits and association with DiD. It is important to reiterate that security is often very simple and intuitive, so always go back to the first principles. Ask yourself why a particular technology was developed and what it solves. A robust defense is not about following the news and trying to implement the latest security buzzword technology in your environment. Every organization is in a unique position with a unique set of assets, and it is very important to assess the cost of an exploit.

In the second part of this chapter, we focused on a particular technology: SOAR. The goal was to infuse a sense of freedom into modern security teams. Gone are the days when **security operations center** (**SOC**) teams would keep an eye on incoming vulnerabilities and detections to take action. Anything that is repetitive needs to be automated to make room for the next problem. Finally, we closed this chapter with the acknowledgment that this field is always evolving and a lot of what works today most likely will not work tomorrow.

In the next part of this book, we will be completely switching our focus to address the evolving threat landscape and explore why DiD works.

Key takeaways

- Security often seems to be a very big problem to solve; divide and conquer!
- Don't buy into buzzwords. Understand the fundamental problem it tries to solve and implement it correctly.
- The zero trust security model encourages us to remove implicit trust and enforce default deny between systems.
- A user, software, or application should only have access to perform its function, nothing more.
- An adequately implemented zero trust security model can trivially contain exposure when it happens.
- Micro-segmentation and securing of each component challenge the traditional stigma of an attacker being right only once.
- Access should be context-aware. Persistent access to critical resources can lead to a false sense of security.
- SOAR technologies can be utilized by every organization to reduce repetitive tasks performed by defenders and have them focus on the next version of the problem.
- Any entity with default trust will be misused by attackers to gain access to organizational systems, including insiders.
- In the world of cybersecurity, everything changes very rapidly and attackers are continuously upgrading their tactics. DiD is the only way to protect. In the case of one control not working, others will still make the attacker's life hard.

In the next chapter, we will shift our focus toward the evolving threat landscape and how DiD becomes the only viable choice for defenders. We will go a little deeper into adaptive defense strategies with a focus on designing defense strategies that can stay resilient for the longer term.

Congratulations on getting this far! Pat yourself on the back. Good job!

Further reading

To learn more about the topics that were covered in this chapter, take a look at the following resources:

- [1] Hunting secrets in public S3 buckets: `https://medium.com/@hareleilon/hunting-after-secrets-accidentally-uploaded-to-public-s3-buckets-7e5bbbb80097`

- [2] Apple's mobile device management: `https://support.apple.com/guide/deployment/intro-to-mdm-profiles-depc0aadd3fe/web`

- [3] Google – Operation Aurora: `https://www.youtube.com/watch?v=przDcQe6n5o`

- [4] Google's official blog on Aurora: `https://googleblog.blogspot.com/2010/01/new-approach-to-china.html`

- [5] BeyondCorp initiative: `https://www.beyondcorp.com/`

- [6] Code owners policy in Chromium: `https://groups.google.com/a/chromium.org/g/chromium-dev/c/aGC8BTTFK64`

Part 3:
Adapting and Evolving with Defense in Depth – The Threat Landscape

In this part, we dive deeper into breaking Defense in Depth further into key elements and present it as an irreplaceable component in any robust defense strategy. The threat landscape underscores that security is never a destination, but an ever-evolving journey. We'll explore emerging threats that demand new defenses, helping you stay ahead of the curve. Recognizing that people are both your last line of defense and often the weakest link, we'll delve into security awareness and training strategies. Finally, we'll solidify the understanding of Defense in Depth as a living, breathing approach.

This part has the following chapters:

- *Chapter 8, Understanding Emerging Threats and Defense in Depth*
- *Chapter 9, The Human Factor – Security Awareness and Training*
- *Chapter 10, Defense in Depth – A Living, Breathing Approach to Security*

8

Understanding Emerging Threats and Defense in Depth

In the dynamic landscape of cybersecurity, organizations constantly face emerging threats that evolve in complexity and sophistication. The emergence of relentless threat vectors, such as sophisticated ransomware operations, state-sponsored attacks, and insidious supply chain compromises, demands a dynamic and ever-evolving approach to defense. This chapter will delve into adaptive defense strategies and illuminate the enduring effectiveness of **Defense-in-Depth (DiD)** principles in today's complex threat environment.

Throughout this chapter, we'll examine various facets of adaptive defense programs, emphasizing the significance of DiD as a fundamental security strategy. A DiD model builds layers of security across your digital assets, from endpoints to networks, applications, and even user identities. While this approach isn't new, its strategic application remains incredibly pertinent. Here, we'll demystify common misconceptions about DiD, showcasing how it works as a cohesive framework rather than merely a collection of tools.

We'll examine why adopting an adaptive mindset within your DiD strategy is vital to combatting emerging threats. From real-time threat intelligence integration to the benefits of automated response actions, we'll discuss how this approach helps minimize overall risk. Additionally, we'll offer insight into evaluating tools and solutions to enhance your existing layers of defense and explore how continuous assessment and modification keep your DiD strategy optimized against contemporary attacks.

Throughout the chapter, practical examples and real-world case studies will illustrate the relevance and effectiveness of DiD strategies in thwarting emerging threats. By the end of this chapter, you will not only have a deeper understanding of the evolving threat landscape but also have learned actionable insights to strengthen your organization's security defenses and ensure long-term resilience against cyber-attacks. Essentially, the chapter will serve as a guide to ensuring your DiD model remains not just robust but agile enough to keep pace with the relentless nature of modern cyber threats.

In this chapter, we're going to cover the following main topics:

- Emerging threat environment
- Adapting DiD to new threats
- Advanced technologies in defense
- Futureproof defense strategy

Let's get started!

Emerging threat environment

The cybersecurity landscape is in a state of perpetual flux. While familiar threats such as phishing and malware persist, increasingly sophisticated actors employ novel attack vectors that challenge conventional defensive strategies. Understanding the dynamics of this evolving threat environment is paramount to recognizing why a simple reliance on traditional DiD models falls short. This section provides an overview of significant new threats, illuminating their disruptive potential and the need for adaptable defense. We'll move beyond generalized risks and highlight specific tactics that expose potential weaknesses within conventional DiD architectures.

This evolution underscores the urgent need to analyze how emerging threats exploit inherent weaknesses in traditional DiD implementations. Assumptions about user awareness, predictable attack paths, and the reliance on static defensive perimeters are continuously challenged by an evolving adversary. As this section examines specific threats, the crucial shift becomes clear: we need to reframe DiD not simply as a series of technological walls, but as a dynamic, integrated strategy that proactively and strategically adapts in tandem with the changing threat landscape.

Evolving ransomware operations

The days of untargeted "smash and grab" ransomware are fading. Ransomware has evolved into a targeted and meticulously planned enterprise, designed to inflict maximum damage and leverage stolen data for multi-pronged extortion. Traditional DiD approaches, which may have focused on endpoint protection and backups, struggle to handle the complexity of these new adversaries.

One alarming tactic is **dwell time** [1]. These sophisticated attacks are no longer about encrypting files at lightning speed. Instead, cybercriminals patiently lurk within a compromised network, quietly mapping critical systems, identifying sensitive data troves, and disabling potential defensive countermeasures. A single-layer DiD, like endpoint protection, is often breached, and this long dwell time lets attackers move unhindered.

One prominent example of dwell time occurred during the devastating 2021 Colonial Pipeline ransomware attack [2]. Here's a simplified timeline illustrating the danger dwell time represents:

1. **Initial compromise**: Attackers used an old, compromised password linked to a dormant company-owned account to gain entry. This highlights the initial perimeter defense failure when outdated accounts persist with access.

2. **Dwell time (over a month)**: Threat actors then spent significant time within Colonial Pipeline's systems. They carefully reconnoitered the network, identified critical control systems for the fuel pipeline, and likely exfiltrated valuable data related to operations and vulnerabilities.

3. **Deployment and impact**: This extensive dwell time gave attackers the detailed knowledge needed to execute a crippling attack on the company's main systems and execute encryption software, shutting down essential fuel supplies to a substantial section of the United States.

The Colonial Pipeline attack highlights how dwell time turns compromised credentials (a seemingly minor flaw) into a devastating and widespread incident. In a traditional DiD approach, detection during this dwell time would have been crucial. Yet, many DiD models lack the tools and continuous monitoring to identify unusual network activity, unauthorized escalation of privileges, and suspicious lateral movement patterns that were likely part of this long-dwell-time attack chain.

Lateral movement is another significant challenge. Rather than solely aiming to encrypt individual machines, adversaries use compromised credentials and exposed vulnerabilities to traverse the network, seeking high-value targets such as backups, databases, and domain controllers. Perimeter-centric DiD models become less effective if threat actors are already inside the network and are free to move laterally.

Finally, data exfiltration transforms ransomware from mere disruption into a catastrophic business risk. Now, simply restoring backups might not be enough if your sensitive data is held hostage or threatened with public release. This underscores the need for defense layers that address not just encryption threats but **Data Loss Prevention (DLP)**, **Identity and Access Management (IAM)**, and exfiltration protection, often overlooked aspects in a more basic DiD implementation.

At a large scale, ownership of software is often the reason for gaps in holistic defense posture. Looking at the Colonial Pipeline ransomware attack from 2021, the defense layers that were missing might sound fundamental in hindsight, but only when we add the complexity of multiple teams owning and protecting different pieces of the software that was exploited do we understand the challenges in building a resilient ecosystem. Using an old password that should have been deleted might seem like an obvious mistake, but let's turn our focus on how getting an initial foothold in an internal network is not that hard anymore.

The rise of deceptive attacks

Deceptive attacks weaponize an organization's most significant strength: *trust*. Deepfake technology and sophisticated social engineering tactics present insidious threats that slip through conventional perimeter defenses and undermine even well-implemented user awareness training. This necessitates a proactive and deceptive approach within a DiD framework.

Deepfakes are prime examples of this danger. Utilizing AI, manipulated audio or video can convincingly impersonate executives, vendors, or other trusted individuals. Requests within deepfakes might bypass perimeter security by asking an employee to initiate actions that compromise defenses, such as disabling two-factor authentication, initiating wire transfers, or sharing sensitive information. These attacks circumvent technical safeguards and prey on established interpersonal trust within an organization.

While deepfakes may be recent, social engineering remains highly effective and continually evolves. Attackers now conduct thorough reconnaissance on social media and company websites. Phishing attacks incorporate personalized language, contextual urgency, and precise targeting that create a convincing guise of authenticity. It's no longer solely poorly worded emails; modern spear-phishing can trick even security-conscious users.

> **Food for thought**
>
> With the rise of Generative AI and the capabilities of large language models, consuming information and building context around unknown data occurs faster than ever. Discuss with your team how this could negatively affect your organization's defense posture.

DiD approaches designed solely around firewalls, user training, and anti-malware measures become woefully inadequate when deception enters the picture. Layering defensive deception methods becomes necessary. Implementing tools for deepfake detection, creating processes to scrutinize urgent requests outside established procedures, and utilizing deception technologies, such as honeytokens or honeypots, can reveal attempted infiltrations that traditional DiD layers might easily miss. This deception element not only detects attacks but also buys time, and it helps gather intelligence on attacker techniques for refining ongoing protection.

Although reported deepfake-assisted breaches are still relatively new, a high-profile case occurred in 2019. Attackers successfully used AI-powered voice mimicry to impersonate the CEO of a UK-based energy firm. They tricked a subordinate employee into transferring €220,000 into a fraudulent account [3]. We will now turn our attention to more AI-fueled attacks.

AI-powered exploits

The threat landscape has evolved to include an arms race of AI. Adversaries leverage ML in a variety of alarming ways, testing traditional DiD models while demanding countermeasures rooted in intelligent, automated adaptation.

Firstly, AI enables unparalleled efficiency in vulnerability scanning. Attackers deploy sophisticated ML-based tools that automate vast network sweeps, rapidly identifying unpatched systems or potentially exploiting configurations. Such AI-powered tools learn over time, refining their vulnerability discovery techniques far beyond static code scanners traditionally used by defenders. A recent research publication demonstrates how LLMs can be used to perform complex injection attacks on websites [4].

Customization of attacks through ML amplifies their danger. Adversaries weaponize ML to build highly targeted phishing emails. Natural language processing techniques generate nuanced text that matches the victim's writing style or incorporates insider lingo specific to an organization. ML analyzes network traffic and user behavior patterns to launch attacks at the times when the system is least monitored, further maximizing impact. These highly personalized, context-aware attacks can slip through conventional spam filters and deceive even security-conscious users.

AI itself can become part of the exploit as ML-based fuzzing software identifies potentially exploitable zero-day vulnerabilities. These tools relentlessly analyze software components, hunting for code anomalies that indicate a path to compromise. Unlike traditional vulnerabilities identified by humans, such weaknesses may evade standard DiD layers, such as firewalls or IDSs, that have built-in protection against known flaws.

Defending against these AI-fueled attacks necessitates the incorporation of intelligence-driven defense layers within a DiD approach. ML-powered behavioral analytics provide a much-needed edge in uncovering subtle patterns and deviations that static rule-based detection may miss. Advanced endpoint protection platforms capable of identifying the behavior of malicious or anomalous code powered by AI become crucial. Implementing threat intelligence feeds that analyze recent AI-powered attacks helps to proactively refine your defensive mechanisms, creating an adaptive shield against new vulnerabilities.

Before moving on to the next section, let's explore a threat category and how AI can power such exploits efficiently.

Automated vulnerability discovery and exploitation

Attackers now develop tools based on ML techniques designed to scan millions of systems at unprecedented speed. These AI-powered vulnerability scanners don't rely on a database of known vulnerabilities. Instead, through techniques such as fuzzing, they learn to identify subtle anomalies in code behavior that reveal potential zero-day vulnerabilities.

Historically exploiting a vulnerability took a long time, particularly because each vulnerability demands expertise in a very core part of computer science. Now this knowledge is much more accessible with the assistance of AI. These examples underscore the speed at which AI-powered scanning tools can target even relatively new vulnerabilities. It challenges the DiD principle of using patch management as a core protection layer. Defense teams must shift toward real-time vulnerability assessment and deploy behavioral analytics along with proactive threat intelligence solutions to detect the signs of such automated AI-led exploitation attempts.

This section has presented you with a lot of challenges that we face today as defenders, to the extent that it might sound impossible to protect our systems. However, in the next section, we will explore how to use the same fundamentals we established in earlier chapters of this book. Keep in mind that the attacks might be new, but we are still trying to protect our assets. Before moving on, let's remind ourselves what needs to change:

- **Speed is critical**: Traditional patch management cycles aren't sufficient against adversaries able to scan, identify, and exploit vulnerabilities rapidly.

- **Context-aware analysis is key**: Detecting unusual patterns from AI-fueled scans (massive numbers of unusual code probes, rapid-fire attempts with subtle variations) becomes a key countermeasure.

- **Threat intelligence**: Feeding DiD with up-to-the-minute data on exploit and AI-scanning techniques empowers defensive systems to detect the "shape" of these novel attacks.

Now let's turn our attention toward why companies need to invest in actively identifying and blocking ongoing reconnaissance attempts; just blocking failed login attempts is not sufficient anymore.

Adapting DiD to new threats

"An ounce of prevention is worth a pound of cure," as famously said by Benjamin Franklin. The cost of exploitation in today's digital world can be grave. It is every organization's top priority today to avoid newspaper headlines for the wrong reasons. Having outlined the pressing nature of emerging threats, it's time to shift focus from analysis to action. Adapting DiD isn't merely about adding new layers or enhancing existing ones; the true evolution lies in building agility and dynamic response into the very fabric of this security philosophy. It's the difference between a fortified castle and a resilient, forward-thinking city capable of responding to unforeseen attacks.

In this section, we'll break down the core tenets of adaptive DiD. We'll discuss how a seamless integration of real-time threat intelligence transforms your defenses from being reactive to proactive. It's about turning intelligence data into automated and intelligent updates to firewall rules, access control policies, and endpoint protection settings. This is more than patching software; it's about using actionable insights to preemptively evolve your defense posture.

The integration of zero trust principles into DiD takes on renewed importance in light of evolving threats. We'll examine how granular access controls, micro-segmentation, and continuous authentication are no longer just add-ons within a DiD model, but critical tools to combat lateral movement and the rising danger of insider threats.

This section is the blueprint for the future of DiD. It's where theory meets practice, and where we redefine resilience not as a static ideal but as instilling a continuously evolving strategy into our defensive DNA, capable of outsmarting even the most dynamic adversaries.

Dynamic risk assessment for prioritization

From the start, we have taken a risk-driven approach (covered in the *Risk-based approach to security* section of *Chapter 1*) and we will continue to do so. The only thing that will change is how we can make our defensive strategy align with the fast-changing threat landscape. Traditional approaches often rely on a static analysis of assets, known vulnerabilities, and generalized threat categories. However, against rapidly evolving attacks, this rigid approach limits agility and response effectiveness. Dynamic assessment shifts the paradigm by prioritizing risks in near real time, empowering you to proactively focus resources where they'll have the most significant defensive impact.

> **Note**
>
> Dynamic risk assessment might seem very confusing at scale, primarily because of changing priorities based on external factors. However, this will vary from one organization to another. Making too many changes too frequently can leave developers frustrated. It is crucial to acknowledge that leaders should come up with high-level directions that abstract the underlying movement. Our recommendation is not to switch projects every week but to use shifting attack vectors as the guiding principle to deliver the highest impact.

At the core of dynamic risk assessments lies the ability to continuously ingest and correlate disparate data. Threat intelligence feeds provide up-to-date analysis of attack vectors, newly disclosed vulnerabilities, and active campaigns targeting your industry. Internal vulnerability scans are contextualized against attack trends; for example, a high-value asset vulnerable to an actively exploited exploit receives higher priority than a known but lesser-impact CVE.

Going further, it incorporates user activity monitoring and network behavioral analysis. Anomalous activities such as unusual login attempts, odd device behavior, or unexpected data flows can be correlated with known active threats. An employee's recent access to highly sensitive data raises the risk profile even when no vulnerability exists, allowing proactive steps such as added scrutiny or enhanced user awareness.

The risk assessment we've discussed so far has been in the context of gathering assets, listing vulnerabilities, and assessing potential loss. We can extend that thinking by associating values to assets across an organization and using context to make automated decisions. One example of this could be detecting unusual activity from an employee's device to a sensitive system while they are on vacation, which could serve as a good indicator to reduce access levels for that device.

While data collection is key, true dynamic nature goes beyond alerts and into quantification. ML models and threat analysis engines help assign dynamic risk scores. These scores consider historical attacks, asset criticality, threat actor activity, and even your past **Incident Response** (**IR**) patterns. This is a growing field and, in a few years, we can expect to see novel innovative approaches to enhance security. As threats, network usage, and assets change, the resulting action adapts accordingly, automatically elevating, and downgrading risk for various DiD layers.

By doing dynamic risk prioritization, we can benefit the entire DiD philosophy. Endpoint protection may lead to real-time prioritization updates based on recent phishing campaigns. Access control is fine-tuned when we can correlate an insider threat with an unusual pattern of resource use. IR is armed with pre-calculated scenarios based on current risk factors – providing agility in a chaotic situation.

Deception-based defenses as a core layer

Throughout this chapter, we have discussed how we need to rethink when detecting adversaries within our boundaries. Defenses to identify anomalous behavior fall short as attackers are now employing high "dwell time" before making their move. Threat intelligence is a good start, but we need to do preemptive threat hunting.

Deception technology disrupts the traditional attacker mindset. Instead of solely focusing on hardening the "real" environment, deception creates a strategically deceptive playground within your DiD infrastructure. This includes honeypots (deceptive systems mirroring critical assets), honeytokens (fake data, credentials, or files), and other lures designed to attract malicious actors.

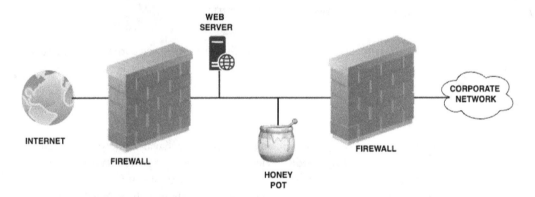

Figure 8.1 – Simple demonstration of honeypots

The core benefit lies in early detection. Unlike perimeter defenses that focus on keeping attackers out, deception thrives on controlled engagement. Interactions with deceptive elements generate high-fidelity alerts, as any activity within these carefully crafted traps definitively implies malicious intent. This minimizes detection delays associated with identifying subtle behavioral anomalies within legitimate production environments.

Moreover, deception becomes a weapon for intelligence gathering. Analyzing attackers' tools and techniques and targeting the deceptive layer provides unparalleled insights into evolving threat trends. Deception isn't simply about detection; it's a source of threat intelligence that continuously enhances your overall defense posture.

It is important to keep in mind that these traps are designed to lure attackers into your environment. Let's break down some key characteristics of effective deceptive defenses:

- **Authenticity**: Good deceptive elements must seamlessly blend into your genuine network, systems, or data stores. Attackers shouldn't suspect they're in a trap, ensuring they interact with them and inadvertently reveal their techniques. This is the most important thing to get right; the challenge is how we do this in a secure way.

- **Targeted**: Tailor your deception around your most critical assets or attack vectors. Placing tempting but too obviously fake traps can reduce their effectiveness, while deceptive systems mirroring your core technology stack hold a higher likelihood of engagement.

- **Interactive**: Allow a degree of interaction to gather deeper intelligence. Simple "tripwire" deception generates mere alerts. More sophisticated honeypots provide limited command access or realistic-looking file structures, letting attackers expose more of their tools and motivations.

- **Dynamic**: Regularly update deceptive elements (files, credentials, and network structures) based on evolving threats and learnings from previous interactions. Stagnant deception traps can quickly diminish their value as attackers learn to identify them.

- **Low maintenance**: A highly effective deception setup should not overwhelm your teams. Go for solutions that don't create excessive false positives or require constant manual tweaking to maintain the illusion of normalcy.

The biggest challenge in deploying honeypots is setting them up correctly. A poorly designed deceptive defense will do more damage than good. Consider seeking expert professional advice if the appropriate skills are not available at your organization.

Now let's turn our focus toward utilizing these non-standard defense techniques and improving our defense. It's one thing to deploy a honeypot and another to effectively drive insights to bolster your defense strategies. Here's how we can make deception actionable:

- **Identify the gaps**: Analyze your existing DiD. Where would early detection of lateral movement or credential theft be most beneficial? Deception placement aligns with those weak spots.

- **Start small, scale smartly**: Low-interaction honeypots or deceptive privileged accounts are quick to deploy, proving value swiftly. As you gain expertise, consider higher-fidelity deceptive networks or specialized deception lures.

- **Integration is key**: Tie deception alerts into your SIEM, SOAR, or EDR for automated responses. Combine these high-fidelity signals with proactive threat-hunting workflows to uncover attacker activity swiftly.

- **Feedback loop**: Regularly analyze attacker interactions with deceptive elements to tailor them more effectively, identify vulnerable assets or processes in your real environment, and update other DiD layers such as access control or user training accordingly.

Remember, deception forces attackers to reveal themselves, giving you the upper hand in a constantly evolving adversarial landscape. It's about turning your defense into a proactive intelligence-gathering machine!

Before moving on to the next section, let's understand the value of this work. First, we need to acknowledge that attack chains sometimes take years to develop. The last thing an attacker wants is to disclose how the exploit was carried out, hence a lot of investment is made in covering their tracks even after successful exploits. The obvious reason behind this is that the same attack will be used against thousands of other organizations. Once an attack chain is published, the security community is generally very efficient at making it widely known. This impacts an attacker's ability to further infect other organizations. When we use deceptive defense methodologies to detect attacker movement in our confined environments early, we have the unique opportunity to gather intel on attacker techniques. This knowledge is invaluable in further thwarting their ability to inflict damage, and additionally, it can act as a deterrent mechanism for attackers.

Next, we will switch gears and explore how we make our IR more efficient. In the previous chapter, we talked about using SOAR platforms to streamline many operations tasks. In the following sections, we'll see how we can use standardization across an organization to make IR smart.

Smart incident response

In the fast-paced battle against evolving threats, static IR playbooks fall dangerously short. Adaptive IR is crucial within a DiD model, emphasizing intelligence-driven agility and proactive lessons learned to improve overall resilience. Playbooks should be continuously updated when newer threats become visible. Let's examine why this shift is so crucial.

Traditional IR playbooks are often built around generalized "if-then" scenarios. While valuable as a foundation, they fail to account for the unpredictable nature of complex and rapidly evolving attacks. Imagine a ransomware incident: a static playbook might initiate system isolation and backup restoration, but what if the threat involves novel lateral movement techniques, or your backups are also compromised? Static IR falters against these unforeseen factors. Adding this additional context to security controls in neighboring regions can drastically increase the efficacy of your defense.

In the previous chapter, we extensively studied how integration with SOAR platforms transforms IR from reactive to context-aware. SOAR's ability to ingest real-time data from threat intelligence, EDR systems, and ongoing investigations creates a dynamic incident picture. Detected anomalies trigger playbook adjustments, quarantines may expand as lateral movement is detected, notification priorities change based on asset criticality or data classification, and additional mitigations kick in if a novel threat tactic is identified. In modern incidence response, SOAR plays a pivotal role.

For IR to be smart, it doesn't stop when the incident is contained. Linking post-investigation learnings back to your defense model is essential to avoid repetitive breaches. If forensics uncover gaps in endpoint protection, those findings get rapidly injected into patch prioritization and software selection. Unusual user behavior observed during an attack drives revisions in security awareness training materials or access control policies.

> **Root Cause Analysis (RCA)**
>
> RCA is a systematic problem-solving methodology designed to identify the underlying cause of a failure, incident, or issue. The goal isn't to merely resolve the immediate problem but to identify the fundamental factors that allowed it to occur in the first place, thus preventing similar incidents in the future.
>
> Many large companies have built a culture around blameless RCA. Thorough RCA during post-incident reviews can be a powerful tool to provide feedback to your system and continuously improve it. RCA reveals the true "gaps" to be addressed. A proper fix extends beyond patch deployment or reprimands. You might modify code review processes, update access controls, or enhance testing to close vulnerabilities at their source, preventing similar future attacks and truly strengthening your DiD posture.

This continuous improvement approach redefines how you view IR. Here's how it changes your outlook:

- **Speed is everything**: Once again, a comment on speed; the faster your IR adapts to an active incident, the better your chances of minimizing impact. Remember, automation enhances speed.

- **Every breach is intelligence**: Don't just focus on recovery. Extract threat data, vulnerability trends, and attacker motives. This is what makes your defense stronger.

- **IR informs defense, and defense informs IR**: Break the siloes; let attack patterns, observed gaps, and emerging threats continuously fine-tune response actions and preventive strategies reciprocally.

One commonly overlooked aspect of IR is standardizing it across an organization. Every incident comes with a sense of urgency attached to it. During the chaos of an active incident, a standardized IR plan promotes streamlined action. Everyone, from junior analysts to incident commanders, follows the same procedures. This eliminates confusion, reduces errors, and speeds up decision-making in stressful, time-sensitive scenarios. A standard template for reporting and escalation ensures relevant stakeholders are informed without ambiguity or delayed communication.

> **RACI matrix [5]**
>
> **RACI** stands for **Responsible, Accountable, Consulted, and Informed**. It's a responsibility assignment matrix often used for IR purposes. A RACI matrix meticulously outlines roles and responsibilities across the diverse stakeholders involved in your IR activities. This includes everyone from security analysts and IT personnel to the CISO and potentially even legal/communications teams. By mapping who is responsible for specific tasks, who has ultimate decision-making authority, who should be consulted, and who is simply kept informed, a RACI matrix ensures streamlined cooperation, reduces the potential for miscommunication, and ultimately facilitates swifter and more effective incident resolution.

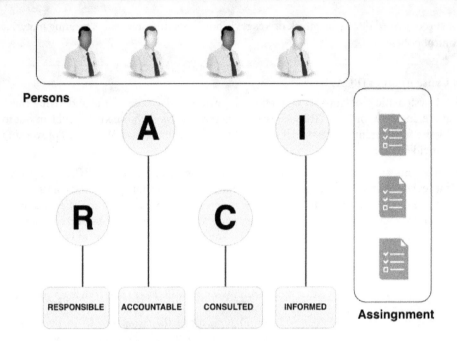

Figure 8.2 – RACI matrix components

As your organization grows, consistent IR policies ease rapid onboarding and training. New security personnel aren't faced with differing approaches depending on their team and there's a foundation they can build upon, minimizing friction during new hire assimilation. This scalability becomes essential when faced with large-scale attacks that need cross-departmental coordination.

Additionally, regulatory requirements demand well-documented IR processes in many industries. Standardized IR policies provide the evidence required and establish transparency for audits. Moreover, they foster clarity regarding ownership and responsibilities, making it clear who should take the lead during various stages of an incident.

Balancing user experience

In this section, we will take a step back and look at our defense controls from a user's lens. Users can be members of the organization interacting with systems internally, partners that often get onto your platform to build products, and so on. Essentially let's understand why making any change very difficult as a defense mechanism is not going to work.

Striking a balance between robust security and positive user experience is a delicate but essential aspect of any effective DiD model. Overly complex restrictions frustrate users, create friction, and lead to unsafe workarounds that worsen your security posture. Conversely, excessive focus on convenience leaves gaping holes that adversaries can exploit.

Adaptive DiD approaches offer a path toward balancing security with usability. It's about intelligently implementing controls that respond to users' authentic needs and behaviors rather than a blanket "strictness" that hampers productivity. Let's explore some examples where adaptive technologies make a real difference.

Transparent MFA is a prime example. Instead of rigid two-factor checks every time, adaptive MFA uses factors such as device recognition, geo-location, and behavioral patterns to adjust authentication frequency. A familiar device on a regular network likely gets streamlined access, while unusual attempts prompt stricter authentication – higher security without constant annoyance.

Similarly, risk-based authentication tailors verification to the context of an action. Logging in to a payroll system triggers stronger factors than routine network access – it's targeted security where it matters most. Session monitoring watches for subtle but risky changes within a session (such as odd location changes or new software used), allowing intervention without constant barriers for ordinary tasks.

Crucially, this balance empowers your workforce as a layer of your adaptive defense posture. Fostering a security-aware culture, where users easily report anomalies and contribute insights without fear of blame, turns your employees into active sensors. Users who understand the "why" of security practices are less likely to circumvent them, and their active observations feed crucial data for adapting controls, threat models, and training.

Remember, in today's complex threat landscape, the "human factor" can't be your weakness; it's potentially your greatest strength. Adaptive defense strategically invests in your people as integral to your defenses, not as a nuisance to manage. We will explore the human factor in defense in more detail in the next chapter.

Emerging tech for the next generation

As we consider the evolving nature of DiD, it's important to have one eye on the future. While thoroughly addressing today's threats is crucial, emerging technologies hold the potential to reshape how we think about our DiD layers in the years to come. Let's briefly explore some promising developments:

- **Homomorphic encryption**: Firstly, homomorphic encryption promises a shift in how we manage sensitive data. Rather than simply encrypting stored data, this technique allows computations to be performed directly on encrypted data itself. This means you could implement defense controls such as threat analytics or anomaly detection without ever decrypting data, reducing exposure risks and potentially facilitating secure outsourced analysis.

- **Blockchain technology**: Blockchain technology offers exciting possibilities for secure asset tracking and immutable logging within our defense models. Tracking software updates, hardware changes, and critical configurations throughout your network environment via an immutable blockchain ledger empowers swift vulnerability identification. Secure, tamper-proof logs create a reliable source of truth for post-incident analysis and compliance processes.

- **AI-powered defense**: Finally, AI-powered autonomous defense looms on the horizon as a controversial but potentially game-changing development. As defensive AI matures, we may see defense layers increasingly capable of predicting, detecting, and even counteracting simple attacks without human intervention. This brings complexity, balancing trust in automated defenders alongside ensuring robust controls to keep this potent technology accountable and aligned with our security aims.

These glimpses of the future underscore the core theme of adaptability. While emerging technologies bring opportunities, their true significance lies in how they're strategically integrated into an ever-evolving DiD philosophy. This leads us naturally into the next section, where we look at the advanced tools and methods reshaping how we protect our digital lives.

Advanced technologies in defense

The previous sections highlighted the necessity for an adaptive DiD approach driven by threat intelligence and responsiveness to dynamic risks. However, to move from theory to practice, we must embrace the cutting-edge technology shaping the tools at our disposal. This section unveils advanced technologies that aren't mere replacements for existing layers but fundamentally transform how a defense strategy functions.

We'll begin with homomorphic encryption. For decades, a trade-off existed – you had to decrypt data to use it but doing so created unavoidable risks. With homomorphic encryption, analysis directly on encrypted data is possible. Imagine systems that process data without ever exposing that data! We'll explore the impact of such a change on data exfiltration protection, cloud environment defense, and the interplay between security and privacy.

We will also explore traditional encryption technologies with forward secrecy to ensure the rise of quantum computers does not lead to our data suddenly being decrypted. Discussions on advanced technologies can't be complete without talking about AI. AI plays a central role in this discussion. AI-powered security solutions move beyond human-written heuristics. We'll address cutting-edge applications in dynamic malware analysis, deep behavior profiling for zero-day threat detection, and AI-driven threat-hunting tools. The balance between the AI advantage and understanding potential biases or vulnerabilities AI systems introduce becomes a key theme in evaluating these potent tools.

On the other hand, we will also look at securing AI itself. Here, we will go over Google's Secure AI Framework and understand how this modern framework relates to the concepts we have been building on throughout the book. Remember, the first principle never changes; we are still protecting our assets! Throughout this section, we'll examine how advances in risk management methodologies fuel modern defense programs. Traditional static risk assessments fail against agile adversaries. We'll discuss context-aware risk management platforms that analyze threat intelligence feeds, internal vulnerability data, and active incident events to generate real-time threat quantifications. As this data flows into automated and adaptable security policies, DiD begins to take on a predictive quality.

Finally, remember that advanced technology needs an equally well-prepared user base. We'll touch on innovative cyber threat training simulations, threat-sharing communities, and fostering cross-sector knowledge exchange – aspects focused on enhancing the human capabilities driving and leveraging cutting-edge defense strategies. Let's begin our exploration of encryption.

Advanced encryption and zero-knowledge techniques

Encryption is probably the oldest technology in the world of security. Encryption's story stretches far back [6], from simple ciphers used by ancient civilizations to complex mathematical algorithms underpinning modern cybersecurity. Early methods such as Caesar's cipher involved simple letter substitutions. World War II spurred the development of electromechanical encryption machines such as the Enigma [7]. The digital age propelled software-based encryption to evolve alongside computing power.

One thing that has always driven progress in encryption algorithms is the ability of an attacker to break them. The strength of a modern encryption algorithm is often correlated with the size of the key. A symmetric encryption that uses a 256-bit key is often more secure than a 128-bit version, and RSA using a 2,048-bit key is stronger than a 1,024-bit version. It always comes down to the attacker's ability to perform complex or lengthy mathematics.

As our computers started becoming increasingly powerful, more and more encryption is being broken gradually. Let's look at how we ensure we protect our information against the rising threat of quantum computers!

Forward secrecy

The looming threat of quantum computers throws a substantial shadow over widely used encryption algorithms. Today's most common schemes, such as RSA and elliptic curve cryptography, rely on mathematical problems that are incredibly difficult for classical computers to solve. This difficulty forms the foundation of our secure online transactions, communications, and data storage.

However, quantum computers operate on fundamentally different principles. Algorithms such as Shor's [8] have the potential to crack these widely used encryption standards with terrifying efficiency. In essence, what would take today's supercomputers thousands of years could be solved by a sufficiently powerful quantum computer in a matter of hours or even minutes. This means an adversary with a quantum computer could potentially intercept secure communications, access sensitive data, and disrupt critical infrastructure relying on current encryption technologies. Although we might be a few years away from economical quantum computers, attackers might be storing encrypted data on their side and waiting for such a revolution in computing to happen. Data generally takes more than a decade to lose its value, and the development of such threats poses serious risks to national security. Beyond that, think about someone having your encrypted credit card information, which is worthless today but, in a few years, if the encryption algorithm is broken, the data is now in the attacker's hands.

Forward secrecy offers a partial but effective defense against the looming threat quantum computers pose to encryption. In traditional encryption schemes, long-term secret keys are used to both encrypt and decrypt data across multiple sessions. Forward secrecy breaks this model, using unique, temporary (ephemeral) session keys generated each time you establish a secure connection.

Suppose an adversary records encrypted traffic today with the aim of deciphering it once quantum computers exist. With forward secrecy, they'll still lack the session keys crucial for unlocking the data. It won't be feasible to derive those past keys even with quantum computing power.

Another key advantage here is that even if a single session key is compromised, only data from that specific session is jeopardized. Past and future communications protected by different ephemeral keys remain secure. It is important to remember that forward secrecy helps manage future risks, but it doesn't prevent real-time compromise if encryption itself is broken. Encrypted communication is still vulnerable to cryptanalysis where adversaries don't need access to the key material.

> **Post-quantum cryptography**
>
> **Post-Quantum Cryptography (PQC)** is a field dedicated to developing encryption algorithms engineered to withstand attacks from a powerful future quantum computer. Unlike forward secrecy, which mitigates the impact of future decryption, PQC focuses on algorithms resilient against the unique computational models quantum machines wield. Current research seeks to identify promising PQC candidates built on mathematical problems that are hard for even quantum computers to tackle. Standardization bodies such as NIST are running extensive PQC competitions to assess their reliability, performance, and suitability for widespread implementation.

Homomorphic encryption

At its core, homomorphic encryption revolutionizes secure data handling. Rather than requiring the decryption of data prior to analysis, it empowers computations to be performed directly on the encrypted data itself. This is made possible through innovative cryptographic techniques that preserve the mathematical structure of the original data even as it is encrypted.

Imagine outsourcing confidential data for processing to a cloud provider. With homomorphic encryption, computations occur without revealing underlying data. The cloud operator sees only encrypted information throughout. This paves the way for privacy-preserving medical analyses, financial calculations, and secure ML as a service.

In addition, sensitive datasets could be shared with researchers or across organizations for collaborative analysis without the need for intricate trust systems or complex de-anonymization. It allows insights to be gleaned without compromising user privacy or ownership of the underlying data.

Types

Homomorphic encryption is primarily divided into three types:

- **Partially Homomorphic Encryption (PHE)**: Supports a limited set of operations on encrypted data (such as addition or multiplication).

- **Somewhat Homomorphic Encryption (SHE)**: Expands operations, but computations introduce noise, limiting overall complexity.

- **Fully Homomorphic Encryption (FHE)**: The holy grail, supporting arbitrary computations on encrypted data, but still computationally expensive for many real-world use cases.

> **Leveled fully homomorphic encryption**
>
> Enables computations on encrypted data, supporting a predetermined depth (number of operations) before accuracy degrades due to computational noise. We did not include it earlier to keep the classification conceptually simple.

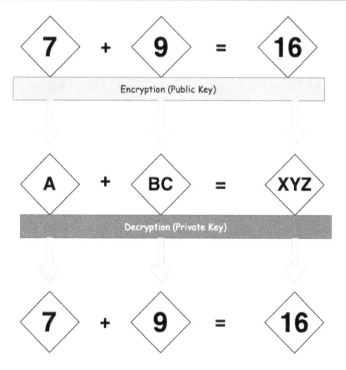

Figure 8.3 – Homomorphic encryption

The biggest challenge with homomorphic encryption is that it demands significant computing power, which could limit real-time applications. Implementing these solutions correctly requires cryptographic expertise and considerations for potential side-channel attacks. As a rapidly evolving field, standardized applications and libraries are still emerging, potentially impacting its mainstream adoption.

Despite the challenges, it is an explosive area of research. Continual advancements in efficiency and standardization point toward encryption at use becoming a key component in next-generation defense architectures. Industries where privacy and security are intertwined, such as healthcare and finance, stand to benefit tremendously from this technology's unique data protection advantages.

Zero-Knowledge Proofs

Another active area of research in cryptography is **Zero-Knowledge Proofs** (**ZKPs**). ZKPs are cryptographic tools that allow someone (the prover) to convince another (the verifier) that they know something without revealing any information beyond the fact that they know it. It's a bit like proving you know the solution to a complex puzzle without showing how you solved it. ZKPs turn knowledge into a form of digital currency that can be exchanged and verified.

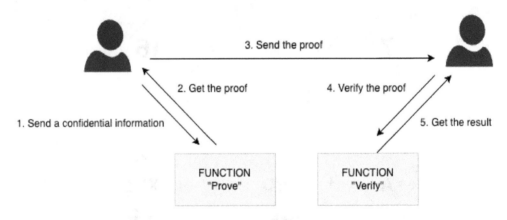

Figure 8.4 – Concept of ZKPs

ZKPs address a fundamental dilemma in digital interactions: how to maintain privacy while guaranteeing authenticity. We frequently need to prove identity, eligibility, or data validity without sacrificing sensitive information. Imagine proving your age without revealing your actual birthdate or demonstrating financial solvency for a loan without disclosing all your account details.

The real core of ZKPs lies in mathematical constructs such as discrete logarithms, polynomial commitments, and hash functions. These allow the prover to manipulate and "commit" to their knowledge in a way that only reveals the truthfulness of the statement, not the underlying knowledge itself.

Three key concepts of ZKPs are as follows:

- **Completeness**: If the prover's statement is true, an honest verifier should be convinced.

- **Soundness**: It should be nearly impossible for a dishonest prover to convince the verifier if their statement is false.

- **Zero-knowledge**: The verifier learns nothing beyond the truth of the statement.

Getting the perfect balance of completeness, soundness, and zero-knowledge is challenging. ZKP construction is a very active area of cryptographic research. The benefits of ZKPs are multi-fold. Here are a few:

- **Enhanced privacy**: ZKPs shield personal information. You can authenticate yourself without revealing underlying data such as passwords, medical records, or financial details.

- **Improved security**: They reduce the risk of data leaks and breaches since sensitive information isn't exchanged in the first place.

- **Scalability**: Due to their compact size, ZKPs can optimize blockchain transactions, increasing speed and efficiency.

Key advancements such as zero proofs, homomorphic encryption, and PQC will have a tremendous impact on how we design our defensive controls. We encourage you to keep an eye on developments in these areas of research. Some of them might be relevant sooner than we anticipate.

Next, we will switch our focus to the biggest technological revolution of the 21st century, the proliferation of AI. We will cover both sides from a defender's perspective: security by AI technologies and security of AI models.

Security by AI

AI and ML hold significant potential in fortifying cybersecurity defenses. These technologies excel at analyzing vast amounts of data, identifying patterns, and detecting anomalies that humans might miss. In the previous chapter, we extensively covered some areas of security where ML is being actively used. Specific ways AI/ML are boosting security include the following:

- **Threat detection**: AI-powered systems continuously monitor network traffic, logs, and endpoints for unusual activity suggestive of breaches. ML models trained on known attack patterns can detect even subtle hints of malware or unauthorized access.

- **Behavioral analytics**: AI can create baselines of normal user and device behavior. Deviations from these baselines raise red flags, enabling the rapid detection of compromised accounts, insider threats, or unusual system activity.

- **Automation and response**: AI aids in automating security responses. Systems can quarantine suspicious files, block malicious IPs, or flag risky actions, freeing up human analysts for more complex tasks.

Utilizing ML in securing our systems is a rapidly evolving area, and sometimes the word AI is thrown around in a security product to make it attractive. A word of caution is essential. AI is not a cybersecurity silver bullet. Just as AI enhances defense, adversaries themselves use it to craft more sophisticated, adaptable attacks. It's important to keep the following in mind:

- **Data bias**: AI models are only as good as the data they're trained on. If that data is incomplete or biased, the models will inherit those biases and fail to detect diverse attack patterns.

- **Adversarial attacks**: Attackers can manipulate AI models with carefully crafted inputs, fooling them into misclassifying threats or misjudging situations.

AI is a powerful tool in the cybersecurity toolkit, but it must be implemented and continuously refined with a critical awareness of its limitations. Its successful use depends on high-quality data, careful evaluation, and constant adaptation to a changing threat landscape. Now that we have looked at security by using AI, let's also explore how we secure AI.

Security of AI

The question of securing AI models and systems is becoming paramount as these technologies become more powerful and embedded within critical sectors. This is a rapidly growing area of research as we implement AI technologies more into critical paths of our lives. We will start by understanding some AI-specific vulnerabilities.

Unlike traditional software, AI models present unique security challenges:

- **Adversarial examples**: As previously mentioned, AI models are susceptible to subtle data manipulations that fool them into misclassifying objects or events. An attacker could carefully alter an image or a line of code to evade detection.

- **Model poisoning**: By feeding malicious data during training, adversaries can degrade a model's performance or inject subtle biases, leading to skewed outputs.

- **Model theft**: Sensitive or proprietary AI models could be stolen to exploit weaknesses or for use by competitors.

As some of these technologies are very powerful, they attract a lot of attention from advanced attackers. There are a few emerging areas of development in securing AI systems:

- **Adversarial training**: To defend against adversarial examples, models are retrained with intentionally perturbed data, increasing their robustness to these attacks.

- **Differential privacy**: This technique introduces controlled noise into datasets during training, making it harder to reverse-engineer private information while preserving a model's utility.

- **Explainable AI (XAI)**: XAI methods provide insight into why an AI model makes certain decisions. This transparency helps debug biases or vulnerabilities.

- **Homomorphic encryption**: As discussed earlier, this is potentially useful when processing sensitive data with an external AI model while keeping the data itself private.

> **Safety and harmfulness**
>
> Safety and harmfulness are very important aspects of developing helpful AI systems. Often known as **responsible AI**, it is a field of study where we implement controls to ensure our AI models do not show bias against specific gender, sex, race, and so on. In this book, we have not covered this topic as it demands a book of its own. [9]

AI security is a rapidly evolving field. As AI advances, we're bound to see more sophisticated solutions and governance models emerge to safeguard this transformative technology. A lot of these defensive techniques are very specific to AI, but our security-first principles don't change. Next, we will quickly cover Google's Secure AI Framework [10] and highlight some of the security mechanisms to better protect our AI systems holistically.

Secure AI Framework (SAIF)

SAIF is a conceptual framework, not a specific tool or technology. It offers guidelines and principles for building AI systems with security and privacy considerations baked in from the ground up. At a high level, the core idea covers the following:

- **Adopt a threat-informed approach**: Proactively identify and model specific threats to AI systems (data poisoning, model theft, and adversarial attacks) throughout the AI development life cycle.

- **Secure and extend best practices**: Build secure foundations by leveraging proven methods from software development (e.g., secure coding and software supply chain security) and extend them to the unique features of AI systems.

- **Detect and respond**: Include AI in your broader threat detection and response capabilities. Train systems and personnel to spot attacks aimed at AI itself.

- **Automate for scalability**: Automate defenses (where possible) to match the speed and scale of AI-powered threats.

- **Harmonize controls**: Maintain consistent security across your AI platforms and environments to avoid gaps in protection.

- **Adapt and iterate**: Embrace an adaptive approach as the threat landscape and AI capabilities change. This implies continually testing and evolving your defenses.

A lot of these pillars are very similar to what we have discussed throughout the book about establishing a robust foundation and adapting. The increasing integration of AI across sectors increases potential vulnerabilities. Addressing these isn't simply about technology, but also processes and collaboration.

Context-aware risk mitigation

So far, we've talked about advancements in security methodologies across the spectrum, starting from encryption to using AI. Now, we will switch gears to go back to one of the fundamental concepts in security, risk management. As we discussed in *Chapter 1*, security is all about balancing a potential loss with business needs. As a result, most businesses today carry a risk registry. Context-aware risk mitigation marks a shift away from conventional security practices that treat every vulnerability listed in a risk registry with equal urgency. Instead, it recognizes that not all vulnerabilities pose the same level of threat. This methodology involves understanding the specific environment, assets, and business implications of vulnerabilities to make smarter risk management decisions.

Key benefits

Context-aware risk management is all about taking input from all sources and contextualizing it with your organization-specific data, with the goal of driving actionable outcomes that are achievable. It helps to do the following:

- **Reduce remediation overload**: Security teams face an overwhelming flow of vulnerabilities. Contextualization helps focus on the most critical weaknesses likely to be exploited, allowing for better use of limited resources.

- **Improve accuracy**: Considering factors such as asset criticality, network exposure, and the presence of compensating controls paints a clearer picture than raw vulnerability scores alone. This leads to more precise risk assessments.

- **Better alignment with business objectives**: By understanding the consequences of a potential breach, security teams can better align their efforts with the organization's overall risk tolerance and business priorities.

Today attackers leverage advanced technologies to exploit vulnerabilities in organizations, resulting in overflowing risk registries. In this landscape, the defender's task often feels like playing catch-up. Given the current scenario, it becomes critical to invest resources effectively in the right areas. Context-aware risk management empowers companies to prioritize the most significant risks.

Futureproof defense strategy

We have been discussing cutting-edge developments in security, but building a futureproof defense strategy requires thinking beyond just chasing the latest technological innovations or reacting to trendy threat headlines. The most resilient architectures rely on a timeless principle: defense in depth. This layered approach offers multiple safeguards against constantly evolving threats and ensures that our defenses can adapt alongside the technology itself.

Fortifying the "layers" for tomorrow involves the following:

- **Network perimeter and beyond**: Perimeter security remains vital, but it evolves alongside cloud migration and remote work trends. Solutions such as zero trust, micro-segmentation, and advanced endpoint protection ensure the perimeter follows the data and user, not just physical office boundaries.

- **Identity as the control plane**: Robust IAM is the new foundation. This involves sophisticated authentication, privileged access controls, and continuous behavioral monitoring to thwart account takeovers that bypass network defenses.

- **Data-centric protection**: Since data is ultimately the target, we need solutions such as context-aware encryption, DLP, and activity monitoring. Protecting data throughout its life cycle is essential, even if a perimeter or device is compromised.

- **Detection and response (reimagined)**: AI-powered analytics and integrated threat intelligence become our eyes, sifting through the ever-increasing noise to prioritize real risks. Automation helps security teams orchestrate and accelerate response, keeping pace with rapidly evolving attacks.

The DiD model excels at providing redundancy and preventing single points of failure. However, futureproof security demands we remain aware that these defenses are only as strong as the people wielding them. Skilled practitioners, user awareness, and a culture of vigilance will bridge the gap between technology and resilience, leaving us better prepared for whatever emerges next.

Summary

The security landscape is in a perpetual state of evolution, advancing in tandem with the identification of new vulnerabilities. In this ongoing dynamic, defenders and attackers play a constant game of cat and mouse. In this chapter, we embarked on a journey to comprehend the ever-changing threat vectors, juxtaposing them against the backdrop of DiD techniques. Our aim was to forge a robust understanding of modern security paradigms.

We delved into common misconceptions surrounding DiD, recognizing it as more than a static set of security controls aimed at thwarting attacks layer by layer. At its core, DiD embodies adaptability. Exploring the concept further, we introduced deceptive security controls as a crucial layer of defense, enabling active monitoring of attacker movements within controlled environments. Additionally, we briefly examined the imperative balance between security and usability, a perennial debate in the field.

Subsequently, we ventured into the realm of advanced security developments encompassing encryption, AI, and risk management. Our intent was to underscore the perpetual nature of the cat-and-mouse chase inherent in cybersecurity. Armed with insights from this chapter, you are equipped not only to adapt your defense strategies using cutting-edge technologies but also to contribute to the advancement of these rapidly evolving research areas.

One aspect intentionally left out of our discussion of emerging threats is humans. In the next chapter, we will focus on the impact that users of a system have on the efficacy of any DiD model. This will provide a complete picture for designing your organization's defense strategy with all components in mind.

Key takeaways

- A core layer of any DiD security model is measuring the efficiency of the controls and continuous improvement.

- Advanced attackers are no longer looking to exfiltrate data the moment they get into your systems. They meticulously observe the environment and wait to maximize their impact without being detected.

- As we get faster and better at consuming information and extracting key insights, traditional multi-layered strategies will fall short. Continuous evaluation of your security posture is key.

- Automated vulnerability discovery and exploits are growing rapidly, making it hard for defenders to keep up with known issues. Speed is critical, and understanding what applies and what doesn't is becoming crucial.

- Keep an eye on recent attack patterns in the industry that your organization is in, and build a defense program that enables a quick shift if/when needed.

- Counter-offence is sometimes the best defense. As a layer of your security posture, consider implementing deceptive defense controls to reveal attackers' identities. Remember, it is very important to get it right.

- Standardize your IR practices. You want to avoid asking questions and looking for resources when the time is not right.

- Keeping an eye out for modern developments in security is important. In addition to teaching what's going on in the industry, it often gives you an idea of the attacker's current tool of choice.

Congratulations on getting this far. Pat yourself on the back! Good job!

Future reading

To learn more about the topics that were covered in this chapter, take a look at the following resources:

- [1] Attacker dwell time: `https://docs.aws.amazon.com/whitepapers/latest/aws-security-incident-response-guide/attacker-dwell-time.html`

- [2] Colonial Pipeline attack: `https://www.cisa.gov/news-events/news/attack-colonial-pipeline-what-weve-learned-what-weve-done-over-past-two-years`

- [3] Deepfake attack story on WSJ: https://www.wsj.com/articles/fraudsters-use-ai-to-mimic-ceos-voice-in-unusual-cybercrime-case-11567157402

- [4] LLM Agents can Autonomously Hack Websites: https://arxiv.org/abs/2402.06664v1

- [5] RACI matrix for ISO 27001: https://advisera.com/27001academy/blog/2018/11/05/raci-matrix-for-iso-27001-implementation-project/

- [6] A Brief History of Cryptography: https://www.redhat.com/en/blog/brief-history-cryptography

- [7] The Enigma machine: https://en.wikipedia.org/wiki/Enigma_machine

- [8] Shor's algorithm explained: https://www.classiq.io/insights/shors-algorithm-explained

- [9] Anthropic Constitutional AI: https://www.constitutional.ai/

- [10] Google's Secure AI Framework: https://blog.google/technology/safety-security/introducing-googles-secure-ai-framework/

The Human Factor – Security Awareness and Training

We started by discussing how humans are often the weak links in our security models, then progressed to the promise of making them an integral part of our defense mechanisms. Throughout this book, we've touched on the idea of reducing the impact of the unpredictability of humans to strengthen our defense strategies. However, we haven't explored practical procedures to realize that idea. Reflecting on large-scale breaches over the past decade, many such attacks began by exploiting this human factor. In some cases, it was through phishing to compromise an employee with extensive access, while in others, it was simply exploiting the security culture of a company.

In this modern digital age, we cannot afford to overlook any known weakness in any part of our ecosystem. Whether you have perfected the implementation of the zero trust model in your security or have inserted multiple defense controls to thwart attackers, humans remain on the critical path to your security success. The simple reason is that these humans carry out tasks that are essential for a company's existence; there's no replacement for them.

The idea of employees clicking their way through the same monotonous security training each year might bring a sigh to many within an organization. Annual security awareness exercises often seem like checkboxes to tick rather than the cornerstone of a truly resilient security posture.

This chapter proposes a fundamental shift in our approach. Forget those tiring training videos – we'll explore how to turn the human element from a potential weak link into an active layer within our **Defense in Depth** (DiD) model. Moving forward, we'll closely examine the human element within security protocols, acknowledging the challenges that are encountered in real-world scenarios.

We'll start our exploration by conceptualizing security as a chain, where each forms a link in safeguarding organizational assets against potential breaches. We'll dissect the unique challenges presented by the human element while examining factors such as distraction and competing priorities that lead to cybersecurity lapses. It's not just about nagging workers; we'll uncover strategies to make security genuinely user-friendly and aligned with productivity goals. This extends beyond basic awareness and into the realm of reliability engineering – how robust systems and secure habits enhance an organization's efficiency in the long run.

Ultimately, this chapter will show you a path toward truly shared responsibility for security. By creating a culture where everyone understands their impact on the security chain, we can empower everyone to take proactive steps in protecting the organization.

In this chapter, we're going to cover the following main topics:

- Security as a chain

- The human element in security

- Security and reliability

- Security is everyone's responsibility

Let's get started!

Security as a chain

In *Chapter 1*, we introduced the analogy of security as a chain. This analogy applied well to traditional security, where defensive controls operated almost independently, and breaking one of them caused the entire system to fail. With the DiD principle guiding modern security models, while a single control bypass might not have as grave an impact as it used to, the core idea about security being a chain is still quite true. In this section, we will explore this ideology with an example.

To understand this analogy, let's revisit what we are trying to achieve with security. While the security bottom line is generally described by the CIA triad (discussed in *Chapter 1*), our mission in security is to prevent unauthorized parties from accessing/modifying our systems/information in an unintended manner. With modern layered security design, one exploit might not grant an attacker access to the entire system, but within our systems, there are ways to establish trust. As the system becomes more complex, these components need to work together to accomplish the task. Essentially, systems need to meet certain sequential criteria to authorize changes and operate. How a payload is delivered to a vulnerable system has become difficult over time, but so have attacker methodologies.

To further dissect this topic, let's focus on the human factor to understand the chain analogy in the modern world. The chain and weakest link analogy perfectly underscores the vulnerability that's exposed by the human factor in security. Even with the most advanced firewalls, encryption protocols, and threat monitoring, a single lapse in judgment or procedure can compromise everything. To illustrate this, let's break down a hypothetical scenario involving privileged administrators and how seemingly small actions can unravel a seemingly secure system.

Consider an administrator with full access to a company's production environment. Due to time constraints, they temporarily disable endpoint protection on their laptop to resolve a software issue. Unfortunately, that device had become infected with malware during a personal browsing session. This breach might go unnoticed initially.

Now, this admin connects to the production system. The malware, which is designed to spread across networks, now penetrates core servers. Data may be encrypted in a ransomware attack, exfiltrated to competitors, or used to cripple entire systems. It all traces back to that initial decision to disable protection on one machine.

Even well-intentioned administrators make mistakes. Perhaps they inadvertently download a malicious attachment believing it's a critical update. Maybe they fall for a phishing attack that hands over their high-level credentials. The point is that even robust security technologies are meaningless if employees with elevated access don't make security-conscious choices at every interaction point.

In a nutshell, the unpredictable nature of humans involved in the loop creates a weak link in an otherwise resilient defense system. One argument could be that a robust security model should not allow humans to access the production environment. While that argument is fair, it is not impossible to leverage any access a developer might have to repeatedly elevate privilege until the malware makes it to production. This can occur via code injection, lateral movement on the corporate network, or privilege escalation on the host laptop by exploiting outdated software, among other methods.

Throughout this chapter, we will explore the human factor from multiple perspectives and learn how we can turn this weakness into a core strength in our security strategy.

The human element in security

People are still at the core of modern software engineering. Humans make decisions, write/maintain software, and operate them to solve business problems. On the other hand, it's humans who attack our systems. The human presence in security is everywhere. If we extend it a little further, it's humans on the other side who are using our services.

We often picture cyberattacks as the work of shadowy hackers lurking in the digital realm. However, some of the biggest vulnerabilities within a modern organization stem from a surprisingly tangible source: the people who run it. From personnel with privileged access to end users juggling emails and work tasks, everyday decisions play a major role in protecting or inadvertently weakening systems.

This section explores the complexities of the human element. We won't be dwelling solely on blame or individual errors; instead, we'll look at real-world challenges faced by various roles within an organization. Why do developers sometimes prioritize quick fixes over security hygiene? How does fatigue play into susceptibility to phishing scams? Can well-intentioned IT admins accidentally break the very chain they aim to protect?

Understanding these factors isn't about finding fault – it's about building empathy. Security fails when it becomes an obstacle to productivity rather than an integrated practice. Today's interconnected tech landscape means securing software isn't just about code quality or network configurations – it's about crafting systems and processes that align with how people work, not how we wish they did.

By the end of this section, you'll better understand why the human element isn't a frustrating workaround to securing technology and that it's a core component that demands the same careful design as any software component. The rest of this section has been carefully divided to help you inspect human factors from different perspectives. We will start by looking into the disadvantages of humans acting on production; then, we will move on to the challenges of developer productivity on not having humans access critical systems, building the stage for an everlasting debate on security and its impact on usability. Finally, we will close this section by looking more into insider risk and how it all fits in a defense program.

Production access

Before going any further, we need to establish what access to production means. In an engineering environment, production access encompasses a wide range of permissions and capabilities that enable individuals to interact with and modify the live, user-facing, influential systems of an organization. Here's a breakdown of the key areas that fall under the umbrella of production access:

- **Infrastructure access**: This includes the ability to manage the underlying hardware and software that run production systems. Examples are configuring servers, databases, networking equipment, load balancers, or cloud resources. Those with infrastructure access can install software updates, change system settings, and have the power to inadvertently interrupt the availability of production services.

- **Code access**: This pertains to the capacity to view, edit, and deploy the actual source code powering applications and services. Developers naturally need this access, but it also comes with risks. Malicious actors or simple errors in production code can lead to vulnerabilities, performance degradation, or even complete system outages.

- **Data access**: Production systems frequently house sensitive customer data, internal business information, or intellectual property. Access to this data is highly regulated and limited to roles with a legitimate need. Data breaches in production environments can have disastrous consequences, both in terms of reputation and legal ramifications.

It's important to note that production access is frequently tiered and role-specific. A system administrator might have broad infrastructure access but no code access. Developers may work primarily at the code level but might need to request temporary elevated privileges to configure settings on production servers for deployment. In the rest of this section, we will focus on individuals with access to one or more of the three highlighted categories.

Role-based access control (RBAC)

RBAC is an authorization mechanism that limits access to resources based on a person's role within an organization. Instead of giving permissions to each user individually, users are assigned roles (such as developer or admin), and those roles have specific permissions. This makes it easier to manage who can do what in a system.

From a security standpoint, the ideal scenario would involve no unauthorized access to an organization's valuable assets, thus mitigating the issue altogether. However, in reality, access is necessary for individuals to carry out their tasks effectively. For instance, an administrator may require production access to troubleshoot an abstract segmentation fault, while developers need access to systems hosting their code.

There are two aspects to this. First, it's the foundation of agility and issue resolution. When developers can promptly access, troubleshoot, and modify production code, they can quickly address bugs, deploy feature updates, and ensure the smooth operation of applications. This ability to respond rapidly boosts customer satisfaction and minimizes service disruptions.

Second, production access is often necessary for monitoring and maintaining system health. Administrators and engineers need access to infrastructure, logs, and performance metrics to detect potential problems, optimize resource usage, and preempt outages. Without it, the organization would be flying blind, leaving critical systems vulnerable to failure and unable to ensure their quality.

Since the inception of this book, we have emphasized that security is a business decision based on risk assessment. Production access represents a significant area where we evaluate trade-offs and determine the level of acceptable risk. Next, let's delve into some of the adverse consequences of allowing individuals to execute arbitrary code in production environments:

- **Increased attack surface**: Production systems are prime targets for hackers. Each individual that's granted access becomes a potential entry point. More access vectors increase the surface area that attackers can exploit through compromised credentials, social engineering, or accidental errors.

- **Accidental damage**: Well-intentioned employees with broad access privileges can unknowingly create vulnerabilities. Misconfigurations, untested updates introduced directly into production, or unintended deletions can lead to outages, data corruption, or security gaps.

- **Insider threats**: While less common, the reality is that malicious insiders with production access have a far greater ability to inflict damage. They can exfiltrate data, plant backdoors, or intentionally sabotage systems, bypassing perimeter security defenses. We will discuss insider risks later in this section.

- **Compliance violations**: Production environments often house regulated data such as financial records, health data, or **personally identifiable information (PII)**. Lax access controls increase the risk of data exposure, leading to significant penalties and reputational damage.

- **Auditability challenges**: Unmonitored or excessively broad production access makes tracking who made what change, and when, incredibly difficult. This hinders investigations of breaches or troubleshooting unforeseen configuration issues.

Often, production access is not a choice but inevitable. However, unmitigated access poses a significant security risk. The goal is striking a balance between enabling employees to do their jobs effectively while implementing rigorous controls to minimize the threats mentioned here. Now, let's shift our focus toward one of the side effects of reducing production access and try to understand the unavoidable human element in security.

Developer productivity

Developer productivity encompasses the speed and efficiency with which developers can design, code, test, deploy, and maintain software solutions that meet business goals. It's not solely measured by lines of code written but rather by the ability to deliver high-quality, functional software consistently while minimizing friction and optimizing resource usage. High developer productivity translates into faster time-to-market, increased innovation, and overall cost savings for the organization.

Reduced production access, while necessary for security, introduces friction into the development process. When developers lack visibility into how their code behaves in the live environment, their ability to identify and resolve issues swiftly decreases. This leads to a chain reaction of negative consequences.

Troubleshooting becomes a bottleneck. Without direct access to logs, errors, or performance data, developers are forced to rely on second-hand information or attempt to recreate issues in staging environments that often don't perfectly mirror production. This increases the time to diagnose bugs, leading to delayed fixes and potential customer dissatisfaction.

The development cycle slows down. The inability to push hotfixes or test small changes directly in production forces reliance on cumbersome deployment processes or approval chains. Experimentation is stifled, making quick iterations a hallmark of agile methodologies, making it difficult to achieve.

Let's consider a real-world example: a critical bug emerges in a payment processing feature on a live e-commerce site. Developers analyze the code but can't pinpoint the root cause without seeing detailed error logs from production. They work on replicating the issue in a staging environment, but that takes hours. Customers experience failed transactions, leading to lost revenue and trust. Finally, with production access, the issue is identified within minutes, resolved, and redeployed, but the damage has been done.

Situations like these are more frequent than we think, and this illustrates that while security is paramount, a complete lockdown on production access impedes developers. The challenge lies in finding a middle ground. Secure access mechanisms and granular visibility to empower developers without compromising security at large. For now, we will pause at the realization that humans are inevitable in production environments.

Security versus usability

Discussing the detrimental effects of reducing production access on developer productivity brings us to the everlasting debate of balancing security with usability. On one hand, software that manages private data cannot be open to everyone for security reasons; on the other hand, users cannot be expected to interact with a system after disabling all other features on their devices. A highly secure application, if not usable for users, will not be widely adopted. Balancing security with usability is a business imperative.

Understanding the eternal tug-of-war

Security and usability often feel like they're locked in an endless battle within the digital world. Security teams strive to build impenetrable fortresses around data and systems, placing strict controls on access and actions. On the other hand, users demand seamless experiences – quick logins, intuitive interfaces, and unhindered access to the information and features they need. Finding the sweet spot between these competing needs can sometimes feel like a mission impossible.

This tension is essential. As we discussed earlier, security safeguards our most valuable assets, including financial details, intellectual property, and customer information. Yet, if security measures become so burdensome that they cripple the actual use of the systems they're meant to protect, they've failed. Similarly, a frictionless experience that leaves data exposed is just as dangerous. It creates a digital environment where people can work efficiently but unknowingly put the organization at tremendous risk.

Security measures and their impact on usability

Now, we will look at some common security controls and their impact on usability to better understand this dilemma. Picture a user's typical workday. To protect sensitive assets, they might face a gauntlet of security hurdles. Complex passwords with frequent expiration dates become a source of frustration, especially when combined with **multi-factor authentication** (**MFA**) [1], which requires phone approvals or hardware tokens. These create additional steps that interrupt workflow and can lead to resentment towards the very measures designed for safety.

Well-intentioned access restrictions are another common obstacle. Even if people have a legitimate need for information or tools, role-based limitations make getting things done tedious. Lengthy approval processes for even minor access changes cause delays, making workers feel hamstrung by security rather than protected by it. These often not only make software less usable but drastically affect developer productivity.

Finally, let's not forget those popups! Frequent alerts about potential phishing emails, certificates, or unknown software can feel intrusive and overwhelming. Users bombarded with warnings face the risk of becoming desensitized, resulting in clicking "accept" habitually, without considering the actual risk. This diminishes the very purpose of those security alerts.

Understanding the difficulty of users regarding security control is an essential element in designing a secure system. Security should not be hard; if it is, people will work around it.

Convenience undermines security

Looking at the same problem from the other side, a system that optimizes for convenience at the cost of everything else is also not very helpful. When security becomes a constant source of frustration, users have a natural human tendency to find ways around it. It may not be malicious intent, but rather a survival mechanism to get their jobs done. Unfortunately, these well-meaning workarounds can have dire security consequences. A complex password policy might lead to passwords being written on sticky notes, shared between colleagues, or overly simplified for easy recall. These behaviors negate the entire purpose of password controls.

Employees struggling with restricted access may resort to **shadow IT**, adopting unauthorized but user-friendly tools to collaborate or share files. This data now lives outside the organization's security perimeter, greatly increasing the risks of breaches and data leaks.

Additionally, the constant barrage of security warnings and interruptions has a desensitizing effect. Users start blindly clicking **OK** or **Continue** on prompts, no longer actively assessing the danger. This renders them vulnerable to real phishing attacks or malware veiled as legitimate popups. The focus on convenience has inadvertently eroded the very security it sought to streamline.

In a nutshell, the conflict between security and usability carries significant consequences. Tipping too far in favor of strict security leads to frustrated users, workarounds that ultimately increase risk, and a general distrust of security measures. On the flip side, prioritizing convenience above all else leaves systems dangerously exposed, putting sensitive data and critical operations at the mercy of attackers.

The path forward lies in a shift toward usability-forward security design. This approach recognizes that security measures can't be built in a vacuum. By understanding user needs, pain points, and workflows, we can integrate security in a way that's intuitive and supports efficient work. This might involve risk-based adaptive security, carefully tailoring controls and alerts, and user-centered designs that make secure behavior the easiest option.

Ultimately, by prioritizing both usability and security, we empower users to become active participants in their organization's defense, not obstacles to it. Throughout the rest of this chapter, we will keep coming back to this core idea. For now, we will switch our focus to another area that significantly contributes to the human element in security models.

Insider threats

So far, we have been debating whether people in an organization should be able to access or modify production assets. Granting production access expands our attack surface so that it includes individuals in positions of authority. Not all hackers are sitting in a remote location; a significant security threat can originate much closer to home, from within the organization itself. Insider threats encompass employees, contractors, or trusted partners who have authorized access to a company's systems and data.

First, we have accidental exposures due to insiders. These well-meaning individuals inadvertently cause breaches due to careless mistakes. Clicking on a phishing link, losing a company laptop, or misconfiguring a system can expose sensitive data or cripple operations. Negligence and a lack of security awareness are major drivers of this risk.

The second type is the malicious insider. Disgruntled employees, those motivated by financial gain, or even those compromised by external actors deliberately misuse their access to sabotage, steal, or leak confidential information. Detecting malicious intent is notoriously difficult as these individuals operate under the guise of legitimate users. It is very common for state actors to have their agents join organizations to launch attack campaigns. Next, we will explore both these scenarios to connect the final dot in the human element in security.

Accidental threat actors

The accidental insider threat often lacks malicious intent but poses a formidable risk. Think of a hurried employee responding to a seemingly urgent email from their "CEO" requesting a sensitive data dump. In the rush of a busy workday, the employee might overlook the suspicious sender address and fulfill the request, inadvertently handing critical information to a cybercriminal.

Falling prey to phishing attacks isn't limited to those unfamiliar with security. Clever social engineering tactics can trick even tech-savvy individuals. A disguised malicious link hidden in a familiar document or a fake software update notification can lure employees into downloading malware or unwittingly giving up their credentials. Running commands on a computer terminal from an "emergency" email from IT services requesting to upgrade a package is more common than you think.

Beyond targeted attacks, everyday oversights can be equally damaging. Accidentally sending confidential documents to the wrong email recipient, leaving sensitive files exposed on an open network share, and losing an unencrypted company device – all of these seemingly minor actions can lead to data breaches with significant fallout.

The challenge with accidental insider threats lies in their sheer unpredictability. Organizations often focus on perimeter security and malicious behavior monitoring, but they need to emphasize ongoing security awareness training and implement systems designed to minimize human error. Sometimes, the weakest link isn't a lack of knowledge but rather the stress, distractions, and sheer volume of information that employees navigate daily.

Malicious insiders

Now, let's shift our focus to the darker side of insider threats: those driven by malicious intent. Unlike accidental insiders, these individuals knowingly exploit their access to inflict harm upon an organization. There are various motivations behind malicious insider actions:

- **Disgruntled employees**: Employees feeling wronged, overlooked, or facing termination pose a substantial threat. They may seek revenge by deleting crucial data, sabotaging systems, or stealing company secrets to take to a competitor. Their familiarity with internal systems makes them dangerous adversaries. [2]

- **Financial gain**: Insider access to sensitive financial information or intellectual property can be incredibly tempting. Individuals might be bribed by external actors or opportunistically decide to siphon data for personal profit. Detecting this type of threat is challenging as the actions can be carefully tailored to mimic legitimate activity. Think about how you control access to your organization's zero-day vulnerabilities – you don't want them to end up in the dark net marketplace.

- **Corporate espionage:** Infiltrators may pose as employees or contractors to gain access to an organization's systems. Their goal is to steal trade secrets, research and development data, or customer information to benefit a competitor or foreign entity. This type of insider threat is often sophisticated and well-planned. State-sponsored attack groups or APTs often take this route to gain valuable insider information so that they can carefully launch their attacks. [3]

It's important to note that malicious insiders often exploit technical vulnerabilities, such as inadequate access controls or logging. However, the root of their ability to operate stems from the trust inherently placed in an "insider." This makes detection exceptionally difficult. Implementing robust zero trust principles, stringent monitoring, and behavioral analytics are essential but not sufficient to uncover malicious insiders before they cause irreparable damage.

In this section, we closely inspected the human element in security. The human element injects an unavoidable level of unpredictability into cybersecurity. Well-meaning employees inadvertently clicking phishing links or state-sponsored attackers wearing employee masks deliberately sabotaging systems underscores the fact that even the most robust defenses can be compromised from within. This makes security a complex business decision; it's never just about buying the latest technology, but about understanding the motivations, behaviors, and vulnerabilities of the people who use that technology.

Security can't be effective if it's seen as an obstacle to getting work done. The next section explores a different lens for security – framing it as an enabler of reliability that empowers developers and fosters productivity. This shift in messaging opens the path for greater security buy-in and creates a powerful alliance between security teams and those building the core infrastructure and software.

Security and reliability

Consider this scenario: a rapidly expanding e-commerce startup prioritizes speed-to-market above all else. To launch new features quickly, developers sometimes take shortcuts in architectural choices. Bypassing security best practices, they quickly deploy and fix issues directly in the production environment. This practice worked perfectly fine for the first few years and made leaders very happy due to the sheer speed at which the organization was rapidly acquiring customers.

A few years down the line, as this company amassed a very large customer base, unexpected events started to happen. An employee who single-handedly designed the architecture was about to leave the company, and suddenly everyone realized that the tribal knowledge was not shared with anyone else. Your first thought upon reading this might have identified this as a security issue. But is it?

The site begins to experience slowdowns and intermittent errors during peak traffic, frustrating developers who had no context on how the systems are operating. A small change to one part of the system triggers unforeseen crashes in a seemingly unrelated area.

As the team scrambles to fix bugs, they find themselves dealing with a tangled mess of poorly documented, tightly coupled code. This means even minor updates take excessive time and introduce the risk of breaking something else. The development team feels trapped in a cycle of reactive firefighting, unable to innovate and deliver the smooth user experience their growth now demands.

Unfortunately, this scenario plays out in countless development teams. The early focus on rapid deployment without a sustainable foundation creates a debt that grows heavier over time – a debt paid in lost reliability, customer satisfaction, and team morale.

While initially appearing as a security concern, it quickly evolved into a reliability and developer productivity issue. This occurrence is more common than you might imagine. Actions such as altering the state of a running pod by copying a patch file from a developer's workstation or restarting a production service by SSH-ing into a machine remain prevalent practices. However, these issues extend beyond security, encompassing broader reliability concerns. In this section, we will explore the connection between security and reliability to present our security improvements as developer incentives rather than productivity costs.

Improving reliability with security

Historically, security conversations often revolve around threats, compliance, and risk mitigation. However, for engineers focused on delivering reliable systems, a different approach can be far more compelling. Often, the most critical security vulnerabilities are also the ones that create unexpected downtime and service disruptions and make it a nightmare to trace and fix production issues.

Let's look at some common security problems through the lens of reliability. We'll uncover how addressing these weaknesses not only strengthens security but also directly contributes to building systems that work as intended, consistently and predictably.

Unmanaged secrets – A reliability time bomb

A lot of the time, we have application keys, database credentials, and API tokens hardcoded, scattered in configuration files, or passed around casually. Developers lack a central system to track and rotate these secrets. These can lead to significant security concerns, but they are bigger reliability problems. Let's understand how:

- **Security issue**: This leaves systems wide open to compromise. If any single secret is stolen, attackers gain a foothold for lateral movement.

- **Reliability nightmare**: One compromised service can cascade to others. Imagine that a database password has been leaked. Due to this, the core data store is inaccessible or corrupted, crippling customer-facing operations. Tracing the breach's origin is a mess, as is revoking credentials across dependent systems without breaking them further.

Failing to properly manage secrets throughout their life cycle can result in security-neutral outages. It's a genuine reliability concern. Imagine a developer setting a unique password for a web server but failing to store it anywhere. Forgetting that secret later could render production inaccessible.

Inconsistent access controls

Production access is granted on an ad hoc basis, without consistent processes or audit logs. Dormant accounts and over-provisioned privileges become common:

- **Security issue:** The attack surface expands unnecessarily, and it's difficult to trace who did what when a breach happens (was it malicious or an accident?).

- **Reliability nightmare:** A developer, trying to debug a production issue quickly, makes an erroneous config change due to excessive permissions. This triggers unanticipated downtime. In their scramble to restore things, the audit trail is muddled, hindering the root cause analysis needed to prevent a recurrence.

Uniformity in managing access control is critical to avoid **access creep** [4]. If ACLs are defined in different places, a production outage due to lack of permission makes it very difficult to remediate the issue. Additionally, these issues are very hard to prevent from recurring.

Unaccounted risk acceptance

Security patching is slow, inconsistent, or skipped entirely to avoid disrupting "working" systems:

- **Security issue:** Each known vulnerability is a welcome mat for attackers, especially for commodity attacks exploiting old bugs.

- **Reliability nightmare:** A ransomware attack locks up critical systems, or a zero-day exploit cripples a core feature. Restoring operations after such incidents is slow, erodes user trust, and proves more disruptive than any planned patching downtime ever would've been.

Engineering teams constantly encounter thousands of vulnerabilities regularly. Due to alert fatigue, developer teams often unknowingly accept highly critical risks. The sheer volume of vulnerabilities often garners the spotlight, but a well-managed risk acceptance program that tracks accountability to individual owners can reduce hours spent in the long run.

These seemingly "just security" problems directly undermine the reliability developers desperately want. It's about unplanned outages, unpredictable breaches, and the inability to quickly diagnose and respond.

In the preceding examples, we discussed common scenarios seen across the industry. Security teams often struggle to garner support from leadership to effectively drive changes across the organization. Securing systems is not as easy as developing the most robust model; it often revolves around convincing humans who are making decisions. Ask yourself, *"What's in it for them?"*

Understanding "what's in it for them"

The truth is that most developers don't prioritize security because they love shipping software fast and efficiently. However, they do value reliability – code that performs predictably, systems designed to handle unexpected loads, and the ability to troubleshoot and resolve issues quickly. Yet, security often feels like a roadblock to this goal – an extra layer of complexity and delay.

This mindset stems from a misunderstanding. Done right, security isn't about hindering progress; it's about illuminating the unknown. Rigorous access controls help you understand exactly what can interact with what in production. Vulnerability scanning shines a light on potential weak points before they become outages. Investing in incident response planning trains the team to restore services rapidly, minimizing downtime.

Think of security as equipping your team with a detailed map of your complex system. It highlights potential hazards, reinforces core structures, and prepares you with alternate routes when the unexpected occurs. This kind of knowledge isn't a burden – it's the key to building systems that inspire trust both from users and the developers responsible for keeping them running smoothly.

Security becomes empowering when framed as an enabler of reliability, not an obstacle to it. From code reviews to reducing access in production, every security milestone comes with a hidden reliability win.

The Illusion of "security versus productivity"

The notion that security inevitably means slower development is a dangerous misconception. The idea is that strict controls add friction, slowing iteration and time-to-market. This ignores the hidden productivity sinkholes that are created when security is an afterthought. Let's consider a few scenarios:

- **The breach aftermath**: A successful attack, whether ransomware or sensitive data exfiltration, grinds development to a halt while all hands are on deck for emergency response. Rebuilding systems, patching, and regaining user trust create massive delays that no streamlined development process could have offset.

- **Death by a thousand outages**: Neglected vulnerabilities or sloppy access controls lead to frequent, unpredictable glitches, crashes, or partial outages. Developers spend their days caught in a reactive loop of troubleshooting instead of focusing on building new features or improving the underlying architecture.

- **The ghost of lost trust**: It's not just about immediate disruptions; security lapses erode user trust. A single well-publicized breach can lead to churn or make it dramatically harder to gain traction in the market, regardless of how innovative your product might be.

Framing the conversation this way reveals that security and productivity aren't enemies but rather interdependent. Neglecting security doesn't make you more agile; it just kicks the can down the road, leading to far more costly productivity losses in the long run.

Benefits of knowing your systems

Picture trying to navigate a vast, interconnected network in the dark. That's what working on complex production systems can feel like without the insights that security tools provide. Robust access controls act like a blueprint, illuminating who (or what service) has access to which components and how data flows between them. This knowledge is invaluable when making changes – you can predict the ripple effects more accurately, reducing the chance of inadvertently breaking an unrelated feature.

Detailed logs and monitoring are like a flashlight in this complex digital landscape. They pinpoint the exact source of an error, trace the chain of events leading to an unexpected slowdown, or help identify potentially malicious activity early on. This level of visibility empowers developers to troubleshoot with surgical precision rather than guesswork, resulting in faster resolution and less disruption.

Think about the anxiety of pushing a code change to production and hoping for the best. Now, imagine having pre-deployment vulnerability scans, automated tests, and staged rollouts backed by real-time monitoring. This security-driven process breeds confidence. You gain a deeper understanding of how changes might impact the system, catching issues before users do.

Incident response planning is another way security boosts productivity. Simply having a plan, knowing who to call, and having backups don't just save time in a crisis – they create a sense of preparedness. This reduces fear-based, rushed decisions under pressure, which reduces the chance of mistakes and prolonged downtime.

At its core, security is about understanding the systems you build. This knowledge shouldn't be seen as restrictive but as a superpower. It allows you to make changes more confidently, respond to problems efficiently, and ultimately deliver systems that not only function as intended but withstand the inevitable challenges along the way.

These benefits are not often immediately obvious to developers, and it is the responsibility of security-minded individuals in an organization to always frame a security problem in a way that makes sense to development teams. Next, we will shift our focus from theories to the practical implementation of best practices that can enhance both the security and reliability of our applications.

Building secure and reliable systems

So far, we've explored how seemingly small security oversights by well-meaning employees or deliberate actions from malicious insiders can lead to data breaches, crippling outages, and long-term loss of trust. Individual training and awareness are vital, but in an environment of constant change and complexity, they're not enough. Relying solely on humans to make the right security decisions at every interaction point is a recipe for inevitable errors and potential exploitation. On top of that, it's also a reliability disaster.

The path forward lies in designing systems that mitigate human risk across the board. This requires a shift away from reactively patching specific vulnerabilities toward proactive architectural principles that embed security and reliability from the ground up. It means eliminating unnecessary points of human access, automating safeguards, and creating well-defined pathways for interaction with critical systems.

The goal of this section is to introduce practical strategies for reducing the attack surface while simultaneously empowering teams to do their work effectively. Imagine a world where production systems are less susceptible to accidental configuration changes, privilege escalation is tightly monitored, and breaches can be contained quickly.

Achieving this doesn't mean eliminating human expertise; it's about enabling people to interact with systems in ways that are both safe and streamlined. This benefits security teams and developers alike, fostering collaboration by aligning their goals toward a more resilient and trustworthy infrastructure. Together, we can create security-focused "well-lit paths" that are easier for developers to build and maintain. Let's explore some common tactics that you can employ to strengthen your DiD strategy, as well as make security easy.

Declarative infrastructure

In traditional infrastructure management, making changes was often a manual, error-prone process. Imagine an engineer logging into a server, updating a configuration file, restarting a service, and hoping they haven't broken anything else. This approach, often termed "imperative," focuses on a series of individual steps to achieve the desired outcome. In the world of cloud providers, a lot of companies rely on engineers logging in to cloud consoles to make changes to IAM policies daily. While this is a good user experience, clicking buttons to change states is hard to track and reproduce.

Declarative infrastructure flips this model on its head. Instead of listing steps, you define the desired end state of your infrastructure in code. Need a database with specific settings? A load balancer with a defined configuration? These are expressed in configuration files (often using formats such as YAML or JSON).

> **Note**
> YAML and JSON are both popular formats for representing data in a human-readable, structured way. They use concepts such as key-value pairs, lists, and nesting to organize information hierarchically. This makes them ideal for defining configuration files, transmitting data between systems, and, as we'll see, for declarative infrastructure tools.

The magic lies in the tooling. Declarative tools such as Terraform [5], CloudFormation [6], and Kubernetes manifests [7] take your "blueprint" and compare it to the actual state of your systems. They then automatically determine the actions needed to align them. If the database doesn't exist, they create one. If the configuration is wrong, they update it. If you accidentally delete a resource, they detect the discrepancy and bring it back.

This declarative approach is like having a tireless robot chef who always makes sure your kitchen is set up exactly as specified by your recipe, regardless of what disruptions might occur. The goal is to reflect your intended configuration to the production environment. Diving deeper into the *hows* of declarative infrastructure, let's divide this into four key components:

- **Configuration languages**: While simple text files can work for small setups, declarative infrastructure thrives on specialized languages or data formats. These include the following:

  - **Domain-specific languages (DSLs)**: Tools such as Terraform use their own language (HCL) designed for **Infrastructure as Code (IaC)**.

  - **Templating**: Systems such as Helm for Kubernetes [8] and Jinja2 allow for more complex logic within configuration files.

  - **Data formats**: YAML and JSON are widely used for their readability and compatibility.

- **State management**: The heart of a declarative system is state management. The tool needs a way to track the desired state (from your code) and compare it to the actual state of the environment. This state information might be stored in a simple file or a database or use APIs to query live information from cloud providers. For example, Terraform can use cloud buckets such as **AWS S3** or **GCP GCS** to store state information.

- **Providers and provisioning**: Declarative tooling doesn't reinvent the wheel. It integrates with "providers" that know how to interact with specific resources, such as AWS instances, Azure resources, database services, and more. When changes are needed, the tool translates your high-level declaration into the API calls or commands required by the target platform.

- **Change automation**: A key aspect is automating changes. The tool analyzes the desired state versus reality and generates a change plan. This includes creating new resources, modifying existing ones, and deleting those no longer needed. Review and approval processes can be built on top of this automation. This is often the hardest part as services such as Terraform provide tools to achieve this but managing the automation becomes very specific.

Now that you understand how to transform your desired infrastructure into code, such as your applications, let's explore the cost of doing this.

Developer perception

Inserting an additional layer between a developer and resources might seem like a loss of developer productivity. Let's look at some examples:

- Fixing a permission issue is much harder than just logging into a UI and updating the IAM policy.

- Adopting new languages, tools, and a declarative mindset requires an initial investment of time and effort.

- Quick iterations by making changes directly to production now get reverted by the control plane when intent skew happens, causing a lot of frustration for developers during the initial adoption.

Changes like these can seem daunting at first, but if properly briefed, they can garner a lot of interest from the developer community. Everyone is trying to do their job efficiently, so highlighting the advantages will increase adoption.

Security and reliability wins

Declarative approaches transform infrastructure management from an ad hoc, error-prone activity into a predictable, auditable process. This has significant implications for both security and reliability:

- Automated provisioning lowers the risk of misconfigurations or accidentally exposed resources that could be exploited by attackers.

- IaC provides a clear history of changes, making it easier to detect anomalies, roll back problematic updates, and comply with audit requirements. The configuration files themselves become a clear, up-to-date representation of your infrastructure.

- In the event of an outage, declarative configurations allow you to quickly rebuild entire environments from a trusted source. From a developer's perspective, provisioning new environments or making changes becomes as simple as updating a code file and letting the tool execute it.

- By treating infrastructure as code, changes are encouraged to happen through the defined process, discouraging on-the-fly tweaks that can undermine stability. This allows developers and operations teams to work from a shared blueprint, reducing misunderstandings.

Overall, declarative infrastructure isn't a security silver bullet, but it significantly reduces the human error factor. This makes it a powerful tool for building systems that are both secure and resilient to the inevitable challenges of production environments.

No human on prod

At its core, the principle of "no human on prod" strives to eliminate direct manual interaction with production environments. It doesn't mean people aren't involved; instead, it emphasizes that human actions should be channeled through strictly controlled processes, tooling, and automation rather than individual access and live changes, as discussed previously.

Imagine a production environment sealed like a high-security vault. Engineers don't log in to servers for configuration tweaks. Database queries happen through predefined interfaces, not ad hoc terminals. Instead, changes are initiated in development environments, expressed as code changes, and flow through an automated pipeline. This might involve code reviews, vulnerability scanning, testing suites, and staged rollouts.

The primary objective is not to make humans obsolete but to minimize the opportunities for unintended consequences arising from manual intervention in a complex live system. A secondary goal is to make every interaction auditable, providing a clear trail of who did what and when, which is crucial for security investigations.

While the ideal is to eliminate direct human access, a pragmatic implementation might involve a "break glass" mechanism, a highly controlled, auditable method for rapid response in true emergencies, and something we will discuss next. Before we proceed, if we were to explore the practical implications of this ideology, it can be broken down into a few core principles:

- **Automation as the default**: The foundation lies in automating every aspect of deployment, configuration, and orchestration possible. This involves the following:

  - **CI/CD pipelines**: Code changes trigger automated testing, security scans, artifact creation, and staged rollouts.

  - **Configuration management**: Tools such as Puppet, Ansible, and Chef enforce desired system states, preventing manual drift.

  - **IaC**: Provisioning and de-provisioning resources happens through versioned infrastructure definitions.

- **Limiting direct access**: Production access controls become extremely strict. These should be implemented with the following principles:

  - **Role-based access**: Only pre-defined roles with minimum necessary permissions are allowed, enforced by tooling.

  - **Temporary privileges**: If elevated access is occasionally needed, it should be time-bound and automatically revoked.

  - **Monitoring and alerting**: Robust logging of access requests and actions on production systems is essential.

- **Shifting workflows left**: "No human on prod" requires re-thinking traditional workflows. This means the following:

  - **Robust testing**: Companies invest heavily in unit, integration, and end-to-end tests to catch issues before deployment.

  - **Staging environments**: Mimicking production as closely as possible becomes paramount for validating changes.

  - **Observability**: In-depth monitoring and telemetry provide the insights needed to troubleshoot remotely.

- **Culture shift**: Success depends on an organization-wide awareness of the importance of limiting direct production access. Developers and operations teams need to collaborate closely, trusting the pipelines and tooling they've built together.

These changes might seem overwhelming at first. Let's understand the perceived productivity loss and how those translate into reliability wins.

Developer perception

The "no human on prod" approach can raise understandable concerns about developer productivity. Let's outline some common perceptions:

- Developers accustomed to pushing fixes directly to prod might feel like pipelines and approval workflows add friction.

- When live access is restricted, debugging production issues may seem more difficult initially.

- Adopting new tools and shifting workflows to "everything as code" requires an investment of time and effort.

- If processes aren't transparent, developers might feel uneasy about changes they don't fully control.

It's important to acknowledge these concerns. Successfully implementing "no human on prod" means addressing them strategically and making sure the benefits to both security and long-term productivity are clearly understood.

Security and reliability wins

Let's reframe those concerns as reliability wins that boost productivity while highlighting their security benefits:

- Automated pipelines, extensive testing, and a consistent deployment process significantly decrease the risk of human error causing production downtime. This translates directly to increased reliability, freeing developers from firefighting mode.

- Knowing that changes pass through a rigorous validation process before reaching production leads to confidence in deploying more frequently, with smaller change sets that are easier to roll back if needed.

- While direct access is gone, rich telemetry, logging, and monitoring provide deeper production insights for troubleshooting. This empowers developers to pinpoint and fix issues remotely more efficiently.

- By minimizing the need for direct production access, the attack surface is reduced. Audit trails become more robust, and access controls are strictly enforced.

Ultimately, "no human on prod" shouldn't be seen as an obstacle, but as an investment. While there's an initial learning curve, the payoff lies in increased system stability, faster problem resolution, and a more streamlined, secure development process.

Access on demand

Earlier, we touched on "break glass" and temporary access. Let's take a closer look at what these terms mean. In traditional models, privileged access to production environments is often granted proactively and persists for extended periods. An administrator might have standing SSH permissions to jump onto servers or broad rights within a cloud management console. While convenient, this leaves a wide window for both accidental errors and malicious exploitation of those credentials.

"Break glass" emergency procedures attempt to mitigate this risk but fall short. They still necessitate privileged accounts "just in case," and the urgency of a crisis can lead to rushed decisions and poor auditing, making it difficult to trace what was changed and by whom.

Multi-party authorization (MPA) [9] adds another layer of security to access control in production environments. It requires approval from multiple individuals before granting elevated production access. This could be a peer, a security team member, or a manager. MPA helps prevent unauthorized actions by a single person and reduces the risk of compromising high-privilege credentials. Combining these two concepts, we arrive at access on demand with MPA.

Access on demand flips the traditional break-glass model. Instead of always-available broad access, it enforces the following principles:

- **Least privilege**: Permissions are only granted for the specific task at hand and at the minimum level required.

- **Just-in-time**: Access is granted temporarily when a legitimate need arises, and automatically expires after a short period.

- **Approval workflow**: Requests for privileged access will go through an approval process based on the sensitivity of the action (MPA).

- **Auditing**: Every request, approval, and action taken within the access window is meticulously logged.

Now that we understand the updated model, let's understand how to implement a similar solution in real life:

- **IAM as the foundation**: Before achieving a seamless access on demand security model, we need to ensure access control is managed in a central place. It can be cloud IAM, a custom company-specific authentication system, or something else:

  - **Centralized IAM systems**: Tools such as Okta, Azure AD, and HashiCorp Vault allow for precise role definitions and just-in-time privilege assignments.

  - **Granular permissions**: Break down permissions to the most granular level necessary (for example, read-only database access versus full administration).

- **Workflow integration**: To ensure smooth interactions with this system, security teams need to ensure they can build automated workflows to make developers' lives easier:

  - **Ticketing systems**: Access requests can be initiated through service desks such as Jira or ServiceNow, linking them to the relevant approval chain.

  - **ChatOps**: Tools such as Slack can integrate approval flows, providing visibility and auditability within developer communication channels.

- **Temporary privilege escalation**: Depending on your team's need, these temporary requests need to be managed either via custom solutions or cloud-native options:

  - **PAM tools**: Solutions such as CyberArk dynamically issue time-boxed credentials that automatically expire.

  - **Cloud-native options**: AWS IAM roles can be assumed for just-in-time access to resources, integrating with approval systems.

- **Auditing and monitoring**: The ability to log and monitor all privileged access activities is crucial. Here, we can implement the following:

  - **Centralized logging**: Every access request, approval, and action taken with escalated privileges must be thoroughly logged.

  - **Alerting**: Set up alerts for suspicious patterns or attempts to use access outside the approved window.

Access on demand is a significant security win as it reduces attack surfaces to smaller windows, but careful thought needs to be put into its implementation. Remember, the goal of any security change should be to make it easy to do the right thing, so implementing all surrounding features becomes as critical as the tool that powers privileged access management. Next, we will explore developers' insights on these changes. However, before that, let's solidify our understanding by going through an example where a developer needs access to a database to debug their program:

1. The developer submits a request with justification and a defined time window.

2. The request goes to their peer/manager and a security team member for approval (MPA).

3. If approved, a PAM tool issues temporary credentials with read-only permissions to the specific database.

4. All activity that was performed using these credentials is logged for future reference.

Developer perception

As we saw, the access on demand model makes significant changes to a developer's workflow, ultimately slowing them down. This shift can certainly raise concerns about potential impacts on developer productivity. Here are some common perceptions:

- The need for approvals and waiting for access might seem to slow down rapid troubleshooting or urgent fixes.

- Strict access controls can be interpreted as a lack of trust in developers' abilities, impacting morale.

- If production access permissions are too restrictive, developers might feel that they are unable to gather the data needed to diagnose issues effectively.

It's important to acknowledge these concerns and proactively address them through transparent processes, well-defined escalation paths, and a focus on equipping developers with excellent observability tools in the production environment. If developers are frequently requesting privileged access, instead of making that process seamless, a better observability solution can be installed to solve the problem at its core.

Security and reliability wins

Access on demand focuses on minimizing risk exposure while enabling developers to get access when truly necessary. This offers several advantages:

- Limiting persistent privileged access minimizes the potential for compromised credentials to be exploited.

- Just-in-time access makes unintended changes to production systems less likely as privileged actions are taken deliberately rather than through always-on capabilities.

- Every access request and action has a clear trail, something that's vital for both compliance and security incident investigations.

- When direct access is controlled, it encourages robust logging, monitoring, and alerting, helping developers remotely understand the system state.

Ultimately, access on demand embodies the principle of zero trust. It reduces the potential for human error to disrupt production while maintaining clear accountability. This aligns security and reliability goals, making systems more resilient to both accidental misconfigurations and targeted attacks.

Reproducibility

Unlike other components we have discussed thus far, reproducibility is something developers actively pursue. In the context of software engineering, reproducibility means the ability to recreate an identical environment, software artifact, or experimental result from the same source code and inputs. Imagine it like baking a cake; if you have the exact recipe (code) and ingredients (dependencies and data), you should get the same cake every time, regardless of who's baking it or what oven they use.

Developers strive for reproducibility for several key reasons:

- **Predictability and confidence**: Knowing that a build process will yield identical results in development, testing, and production instills confidence that changes are unlikely to introduce unexpected side effects. This enables faster release cycles.

- **Simplified debugging**: The ability to reproduce a bug locally on a developer's machine, mirroring the production environment, is invaluable for troubleshooting. It eliminates the guesswork of "it works on my machine" scenarios.

- **Ease of collaboration**: Reproducible setups enable seamless sharing of work between team members and across different locations. New developers can get up and running quickly without wrestling with environment-specific configuration issues.

- **Disaster recovery**: In the event of an outage, being able to replicate the production environment rapidly minimizes downtime and ensures a consistent recovery point.

The pursuit of reproducibility reflects a desire for control over the complex interplay of code, dependencies, configurations, and infrastructure that form modern software systems. It ultimately allows developers to focus more on building features and less on battling inconsistencies in their environments.

Security and reliability wins

Let's examine why reproducibility acts as a powerful force multiplier for both security and reliability:

- Reproducible environments inherently minimize unintended variations across development, testing, and production environments. This translates to fewer exploitable discrepancies that attackers might leverage. Unpatched vulnerabilities are applied uniformly, leaving no gaps in protection.

- The ability to replicate issues in isolated environments allows for rapid testing of patches and security workarounds without risking further disruption to production. This means vulnerabilities can be addressed with speed and certainty, reducing the window of exposure.

- When deployments and rollbacks become predictable due to reproducibility, it acts as a deterrent against malicious actors. Attackers often rely on complexity and unexpected system behavior to their advantage. Reliable processes make their job significantly harder.

- Reproducing the environment at the time of a security breach provides deeper insights into the attacker's methods. This invaluable forensic data allows for building stronger defenses and preventing similar attacks in the future.

> **Hermetic builds**
>
> A hermetic build refers to a self-contained build process where all dependencies and configurations are explicitly defined and bundled. This means the build can be reproduced identically, regardless of the machine or environment it runs on. Hermeticity is a foundational principle for reproducibility, ensuring that code behaves consistently throughout development, testing, and production, contributing to both security and reliability.

Reproducibility provides a shared language and framework for developers and security teams to collaborate effectively. Ensuring environments and processes can be consistently replicated reduces friction caused by remote debuggability and unpredictable deployments. This predictability empowers developers to proactively integrate security checks early in the development cycle while giving security teams confidence that vulnerabilities will be addressed consistently across the entire system. Ultimately, reproducibility fosters trust and alignment, transforming security from a perceived roadblock into a shared responsibility that strengthens reliability for everyone.

Security is everyone's responsibility

Traditionally, security has often been viewed as a separate department, a team responsible for firewalls, compliance checklists, and saying "no" to developers in the name of risk mitigation. This approach creates an adversarial dynamic, pitting security against innovation and ultimately hindering the organization's very mission. The reality is that in a world where data breaches and cyber threats constantly evolve, security can't be an afterthought or a bottleneck; it needs to be a fundamental part of how every system is designed, built, and operated.

The human element, as we've explored, is a double-edged sword in security. While individuals can be the weakest link, they can also be the most powerful line of defense. The human element is the most commonly overlooked area in an otherwise robust DiD strategy. The goal of this section is to reframe security as a shared responsibility. It's about empowering every engineer, designer, and product manager to proactively consider the safety and privacy of the user data entrusted to them. This isn't about adding extra hurdles; it's about embedding security principles into the process so that they become second nature.

Let's explore how to break down silos and foster a security-conscious culture. We'll address common obstacles faced by security teams and strategies to reposition security as an enabler, a vital partner in delivering reliable, trustworthy products that users can confidently rely upon.

Common challenges security teams face

Security professionals play a vital role in safeguarding an organization's digital assets and protecting sensitive user data. However, their path is rarely smooth. Beyond the technical complexities of defending against sophisticated threats, security teams regularly encounter internal roadblocks that hinder their ability to make meaningful improvements.

Understanding these challenges is the first step toward bridging the gap between security goals and the everyday realities of development and business operations. These obstacles range from competing priorities and communication breakdowns to the complexities created by the rapid pace of technology adoption.

It's important to acknowledge that these hurdles aren't born from malicious intent. Rather, they highlight the need to reframe how security is perceived and integrated into an organization's DNA. Let's delve into some of the common pain points security teams encounter, setting the stage for future discussions on how to overcome them:

- **Competing priorities**: Security initiatives often clash with the relentless drive for feature releases and short development cycles. Developers might view security measures as slowing them down, while business leaders may be hesitant to invest resources in something that doesn't directly and immediately generate revenue.

- **Communication gaps**: Security teams often struggle to communicate risks and vulnerabilities in a way that resonates with non-technical stakeholders. Using overly technical jargon or focusing solely on doom-and-gloom scenarios can lead to disengagement and a lack of urgency.

- **Resistance to change**: Introducing new processes, tools, or security controls can be met with resistance. People are naturally inclined to stick with familiar workflows, and security can be misconstrued as simply adding hurdles and bureaucracy.

- **Resource constraints**: Security teams are frequently understaffed and overworked. This makes it difficult to proactively address vulnerabilities, stay on top of the evolving threat landscape, and effectively advocate for necessary security investments.

- **The "blame game"**: When breaches or security incidents occur, a culture of finger-pointing can emerge, pitting security teams against development and operations. This hinders collaboration and makes it harder to learn from mistakes. It also erodes trust and makes security teams defensive, hindering future progress. Security teams often rely on inserting them in "approving" changes without appropriate resource planning.

> **Note**
> If security teams are not properly staffed, they should not make decisions for development teams by approving or denying changes. This sets the stage for developing a hostile relationship. Instead, security teams should aim to position themselves as advisors to help make the right decisions. Remember, policies mean nothing without enforcement. If security cannot verify whether their advice is being followed, there's not much to be gained from simply "denying" changes.

- **Alert fatigue**: Security tools can generate overwhelming amounts of alerts, many of which turn out to be false positives. This desensitizes teams to the point where critical warnings might be overlooked amid the noise. A very common scenario is vulnerability scanners generating thousands of findings every day. It is the security team's responsibility to address that unending list and only pass on "true positives" to developers.

The landscape of challenges faced by security teams underscores the complexities of safeguarding digital assets within organizations. Success hinges on more than just technical expertise. Overcoming silo mentalities, fostering clear communication, and proactively addressing resource constraints are equally crucial. Both security teams and their development counterparts must strive to reimagine processes, embracing collaboration and shared responsibility to create a truly secure environment.

Now, let's turn our attention toward the changes a security team can bring about.

Your security toolkit

While technology plays a crucial role in security, true success goes beyond firewalls and intrusion detection systems. Fostering a culture of security awareness and collaboration requires a multifaceted approach. This involves making security tangible for non-specialist team members, aligning security goals with the overall business mission, and equipping developers with the tools to embrace secure practices without undue friction.

Think of the following tools and strategies to bridge the gap between security teams and the rest of the organization. They're designed to spark dialog, build empathy, and demonstrate the practical value of prioritizing security in every stage of product development.

Let's explore some approaches that your security team can leverage to break down silos, gain buy-in, and foster a culture where security is everyone's shared responsibility:

- **Gamifying security**: Capture attention and drive awareness through interactive means. **Capture the flag** (CTF) exercises, simulated breach scenarios, and even gamifying secure coding practices can make security concepts tangible and engaging for both technical and non-technical team members.

- **The power of storytelling**: Move beyond raw statistics and technical jargon. Share real-world examples of security breaches relevant to your industry, translating their impact into lost revenue, damaged reputation, or compromised customer trust. This drives home the "why" behind security efforts.

- **Metrics that matter**: Shift the focus from simply tracking the number of vulnerabilities patched to highlighting the tangible benefits of security improvements. Emphasize metrics such as reduction in security-related downtime, faster incident response, or the proactive mitigation of high-impact threats. This connects security work to core business objectives. Some security adjacent metrics such as reliability gains are also valuable to establish a common ground with development teams.

- **Threat modeling workshops**: Make threat modeling a collaborative exercise that involves developers, product managers, and security. This proactive approach builds a shared understanding of risks and surfaces potential security weaknesses early in the design phase, making fixes less disruptive and costly.

- **Frictionless DevSecOps**: Invest in tools and processes that seamlessly integrate security into the developer workflow. Automated code scanning (SAST, DAST), vulnerability feeds tailored to the technologies you use, and clear remediation guidance empower developers to take ownership of security without feeling like it's an adversarial process.

- **Celebrating success**: Publicly recognize security wins, proactive fixes found through security champions programs, and successful collaborations between teams. This reinforces positive behaviors and builds a culture where security is viewed as a contributor to success, not an obstacle.

Remember, security is an ongoing journey, not a destination. Consistent use of these tools, coupled with open communication and a willingness to adapt, will go a long way in transforming security from a perceived roadblock into an integral part of your organization's success.

Summary

Humans are the unpredictable element in any defense strategy, capable of both bolstering and undermining its effectiveness. Their involvement in creating, maintaining, and operating systems introduces an inherent vulnerability. This chapter explored the human factor in security, aiming to harness it for stronger defenses. Understanding human behavior is key to anticipating the unexpected and building awareness. We need to move beyond the traditional approve/deny security model that often hinders innovation.

Security teams must shift their perspective to effectively support efficient development. Security needs to become a transparent, seamless service that's integrated into the development process. By focusing on metrics, developers can prioritize things such as reliability, availability, and reproducibility – security teams can create intuitive, "well-lit" development paths. This proactive approach fosters a security-conscious mindset across the organization.

This mindset shift aligns with DevSecOps principles and fosters a stronger security culture across the organization. The key is collaboration: security teams must offer actionable guidance and create clear pathways for secure development. This approach not only improves security but also boosts efficiency and overall company culture. You are highly encouraged to consider how these concepts apply to their organizations and identify potential challenges and solutions for successful implementation.

Throughout this book, we have examined the DiD model from various perspectives, beginning with fundamental concepts and extending to the unpredictable human elements in security. In the final chapter of this book, we will explore how DiD emerges as the sole viable security strategy. We will delve into newer frameworks such as SSDF and assess their impact on the security landscape of tomorrow.

Key takeaways

- The human element might be the weakest link in your security model today, but it doesn't have to be.

- Focus on building a shared security mindset across a company. This will contribute to bigger security wins than installing a new security tool.

- Security controls that hinder developer productivity are generally not very effective as teams tend to work around them.

- Reliability is security's biggest friend that gets recognized by developers.

- A right balance between security and usability needs to be implemented.

- Modifying the production environment is almost always a bad idea, not just for security.

- Before proposing any security changes to leadership, focus on one question: what's in it for them? Your conversations will go much smoother.

- Descriptive infrastructure is not just a security improvement – it's a developer win as well.

- To master DiD, security teams need to invest in communication as much as the latest technologies.

Congratulations on getting this far – pat yourself on the back! Good job!

Further reading

To learn more about the topics that were covered in this chapter, take a look at the following resources:

- [1] What is MFA?: https://aws.amazon.com/what-is/mfa/

- [2] Statistics on Disgruntled Employees: https://www.informationweek.com/cyber-resilience/75-of-insider-cyber-attacks-are-the-work-of-disgruntled-ex-employees-report

- [3] National Insider Threat Task Force: https://www.dni.gov/index.php/ncsc-how-we-work/ncsc-nittf

- [4] What is Privilege Creep?: https://heimdalsecurity.com/blog/what-is-privilege-creep-and-how-to-prevent-it/

- [5] Terraform, IaC: https://www.terraform.io/use-cases/infrastructure-as-code

- [6] AWS CloudFormation: https://aws.amazon.com/cloudformation/

- [7] Kubernetes Resources: https://kubernetes.io/docs/concepts/cluster-administration/manage-deployment/

- [8] Quick start with Helm: https://helm.sh/docs/intro/quickstart/

- [9] Multi-Party Authorization: https://en.wikipedia.org/wiki/Multi-party_authorization

- [10] Google Insider Guide to Building Systems: https://sre.google/books/building-secure-reliable-systems/

Defense in Depth – A Living, Breathing Approach to Security

The pursuit of absolute security is an illusion. In a world where cyber threats constantly evolve and new vulnerabilities emerge, we can never fully eliminate risk. However, this doesn't mean we're helpless. The goal shifts from aiming for an impenetrable fortress to building systems that are resilient, and where breaches are difficult, costly, and time-consuming for attackers.

This is where **Defense in Depth (DiD)** comes into play. It's a philosophy and a collection of best practices that acknowledge the inevitability of individual security layer failures. DiD creates multiple, overlapping lines of defense throughout a system, making it significantly harder for an attacker to succeed, even if they manage to breach an initial layer. Throughout this chapter, we'll demonstrate why DiD has become indispensable to any modern security model.

We began by establishing the fundamentals of security and DiD in the earlier part of the book, then shifted our focus to examining problems from an attacker's perspective to refine our defenses. Thus far, our emphasis has been on developing a deep understanding of the individual components that constitute a comprehensive defense strategy. In this final chapter, our objective is to translate those insights into a practical implementation that can assist us in constructing a resilient security posture that is prepared for tomorrow's attackers.

Security isn't a destination you reach and then rest on your laurels. It's an ongoing adaptation to the relentless evolution of threats and the ever-expanding complexity of the systems we build. In this chapter, we'll challenge the notion of "bulletproof" software, acknowledging that vulnerabilities are inevitable as technology advances.

Our goal is to build a DiD mindset, utilizing the **Secure Software Development Framework (SSDF)** as an illustrative example for building multi-layered defenses. We'll stress the importance of continuous monitoring and improvement once again, emphasizing that security isn't a project with an end date, but rather a mindset that must be integrated into every stage of a system's life cycle. Finally, we'll underscore why only a "living" DiD model, one that constantly adapts, can hope to keep pace with the sophisticated attacks of tomorrow.

In this chapter, we're going to cover the following main topics:

- Security is relative
- Operationalizing DiD with the SSDF
- Continuously monitoring and improving security posture
- Security tomorrow – Sustaining a living DiD

Let's get started!

Security is relative

The pursuit of absolute, unbreachable security is a fool's errand. In a world where attackers relentlessly refine their techniques, new technologies expose unforeseen vulnerabilities, and the sheer complexity of modern systems outpaces our ability to fully test them, perfection is unattainable. To assume your software is impervious is to invite disaster.

Attackers are evolving at an alarming pace. Sophisticated AI tools automate the discovery of vulnerabilities [1], tirelessly probing for the tiniest cracks in your defenses. Each zero-day exploit discovered and each credential leaked gives them an edge. We are in an escalating arms race, and complacency is not an option.

The key to survival isn't clinging to the latest technology at your defense. It lies in understanding that failure at some level is inevitable. True resilience comes from designing systems with the expectation that individual components will be compromised.

This mindset is the foundation of DiD. By creating multiple, overlapping layers of security, we force attackers to expend far more time and resources, increasing their chance of being detected or stymied before they reach their ultimate target. While we can never fully eliminate risk, we can dramatically tilt the odds in our favor.

In *Chapter 1*, we introduced the concept of a **reference monitor**. The idea was to minimize the attack surface by exerting additional efforts to ensure that software performs precisely as intended, and nothing beyond that. The key principle was to keep it small enough to be verifiable. However, in modern-day systems, this principle does not hold true. Let's explore some characteristics of modern systems that make achieving absolute security unattainable.

The complexity factor

The software systems we rely on today are intricate webs of interconnected components. Microservices architectures, cloud-based infrastructure, third-party libraries, and legacy code bases all come together to form a tapestry that's often mind-bogglingly complex. While this complexity enables amazing functionality, it also creates a breeding ground for unforeseen vulnerabilities.

Consider a simple web application. A seemingly innocuous change to a frontend component might inadvertently cascade to expose a misconfiguration in a downstream API, creating an opening that wasn't there before. The sheer number of moving parts, potentially owned by different teams, makes it impossible to track every possible interaction and guarantee that no unintended side effects emerge.

Complexity also obscures visibility. Understanding the true attack surface of a system becomes a monumental task. Shadow IT, forgotten test environments, or undocumented dependencies create blind spots that attackers eagerly exploit. It's the digital equivalent of a sprawling mansion with countless hidden entrances.

Moreover, complexity breeds configuration drift. The ideal secure state, as defined when a system is designed, gradually erodes over time. Quick fixes, workarounds, and updates can leave systems in a Frankenstein-like state [2] where security assumptions no longer hold true.

It's a sobering reality: the very complexity that allows us to innovate and deliver exceptional user experiences simultaneously increases the difficulty of securing those experiences. DiD, as we've explored throughout this book, is an essential response to this challenge.

Legacy systems

Legacy systems are the ghosts that haunt modern infrastructure. These are applications built on outdated technologies, often maintained by a dwindling pool of experts, and running on hardware that may no longer be supported. While critical to business operations, they represent ticking security time bombs.

The core problem is that legacy systems were often born in a time when cybersecurity was an afterthought, not baked into the design. They may lack fundamental security features we take for granted, such as input sanitization, robust authentication, or memory protections. This makes them inherently more susceptible to entire classes of common attacks.

Making matters worse, patching becomes problematic. Vendors may no longer provide security updates for older operating systems or software frameworks. Even if patches exist, applying them risks creating unpredictable compatibility issues that can break the entire system. Organizations are forced to choose between stability and security.

The difficulty of hardening legacy systems extends beyond the code itself. Documentation may be sparse or non-existent, making it hard to understand how the system interacts with newer environments. The original developers may have moved on, leaving a knowledge vacuum.

The unfortunate reality is that legacy systems can act as "soft targets" for attackers. They know the vulnerabilities are there, and they understand defenders are often reluctant to make drastic changes due to the risk of disruption. This persistence puts pressure on every other layer of defense, as a breach in the forgotten legacy corner could give attackers a foothold to pivot toward more sensitive assets.

Complex and side-channel attacks

Side-channel attacks take an unconventional approach. Instead of directly targeting bugs in code, they focus on unintentional "leaks": subtle changes in a system's timing, power consumption, electromagnetic emissions, or even sound. These seemingly insignificant signals can be analyzed to reveal sensitive information or bypass security controls.

Imagine a web application that diligently protects against SQL injection attacks. Its input validation is very exhaustive. However, a side-channel attack could potentially still deduce information about the database structure. By carefully measuring how long different queries take to execute, even when they return no results, an attacker might be able to infer whether a table or column with a specific name exists. This knowledge becomes a piece of a larger attack puzzle.

Blind SQL injection [3] specifically highlights the insidious nature of side-channel attacks. While traditional SQL injection relies on getting the application to directly reveal data through error messages, a blind attack uses true/false deductions, such as timing differences, to painstakingly reconstruct data one bit at a time. It is debatable whether blind SQL is a side-channel attack, but the idea here is to demonstrate how some vulnerabilities in a system are very hard to identify.

Side-channel attacks are a sobering reminder that security isn't just about the code we intentionally write. The underlying physics of how computers operate creates vulnerabilities that are incredibly difficult to fully defend against. They demand vigilance not just in software design, but also in the choice of hardware, deployment environments, and constant monitoring for anomalies that might signal this type of attack.

This emphasizes why aiming for perfect, unbreakable security is unrealistic. Even the most secure software might fall prey to an attack exploiting the very laws of physics it runs on. Let's quickly look at a case study of a recent side-channel attack.

Spectre and Meltdown attacks

Spectre and Meltdown [4] are a class of vulnerabilities that sent shockwaves through the tech industry a few years back. They fundamentally leverage the way modern processors optimize performance through something called "speculative execution" [5]. Here's a simplified breakdown:

- **Speculative execution**: To speed things up, processors often guess what instructions a program will need next and execute them ahead of time. If the guess is right, performance improves. If it's wrong, the results are discarded.

- **The Spectre/Meltdown exploit**: These attacks trick the processor into speculatively executing instructions that leak sensitive data from protected memory areas. This data is then subtly encoded into the processor's cache state.

- **Reading the cache**: Through meticulous timing analysis, a malicious program can deduce the leaked data from these cache changes, even if it doesn't have direct memory access. Think of it like trying to figure out the shape of an object by the way it displaces water in a container.

Spectre and Meltdown were devastating because they exposed secrets thought to be isolated: passwords in a browser's memory, encryption keys on cloud servers, and so on. They required patches to operating systems and browsers and even firmware updates to mitigate.

What makes Spectre/Meltdown a chilling reminder is that they exploit the fundamental quest for performance in processor design. While mitigations exist, similar side-channel attack techniques continue to emerge, demanding ongoing vigilance. This is a humble reminder that as we upgrade our defense stack, attackers will adopt newer techniques to break the defensive barrier. With DiD, we can ensure that each layer provides enough deterrence for attackers to make their attacks unsuccessful.

Now, we will use a framework to understand how to effectively operationalize security into your SDLC.

Operationalizing DiD with the SSDF

We've acknowledged that the pursuit of absolute security is a futile mission. Vulnerabilities, evolving threats, and the complexity of our systems are forces we can't fully control. This might seem disheartening, but it's actually empowering. It frees us from the illusion of perfection and propels us toward strategies designed for resilience in a world of constant risk.

DiD is one such strategy that has been the focus of this book. It embraces the inevitability of individual failures and focuses on building layers of overlapping protection that force attackers to overcome multiple hurdles. But how do we move from theory to practice?

The SSDF [6] provides a structured approach to operationalize DiD principles. It offers guidance for integrating security at each phase of the development life cycle, from initial design to ongoing maintenance and incident response.

In this section, we'll dissect the SSDF, examining its key phases and the practical security controls it recommends. Think of it as a blueprint for building software that's not only functional but also built to withstand the relentless assaults of the modern cyber threat landscape. Our goal is to equip you with the knowledge to operationalize a security model based on a security standard that applies to your organization.

Understanding the SSDF

The SSDF, as its name implies, is a comprehensive set of best practices and guidelines designed to bake security into every stage of the software development process. Developed by security experts at NIST, it recognizes that security isn't an afterthought bolted on at the end but rather an integral quality characteristic built into the software from day one. The SSDF was created in response to President Biden's Executive Order 14028 [7]. It encourages you to think like an attacker from the very beginning of a software project. What valuable assets will your system handle? How might someone try to steal or disrupt them? By considering threats upfront, you can design defenses directly into the architecture, avoiding costly bolt-on security later.

The SSDF provides guidance on writing safe code that's less likely to have those accidental vulnerabilities that attackers love. This includes training developers on common attack types, providing secure coding libraries, and automatically scanning code for problems as it's being written.

The SSDF goes beyond protecting just the software itself. It recommends hardening the operating systems it runs on, locking down network configurations, and carefully controlling access to reduce the "blast radius" if something does go wrong. This layered approach makes it harder for attackers to move around inside your systems if they breach an initial defense.

This framework recognizes that even the best-built systems need watching. Logging, security monitoring tools, and regular vulnerability scans help us spot attacks early or uncover new weaknesses that have appeared over time. This emphasis on "assume breach" means being ready to react quickly and minimize damage.

Individually, these are some concepts we have repeatedly gone over throughout the book. Think of the SSDF as a battle-tested blueprint for building resilient software. It combines time-tested techniques and proactive thinking to help your systems withstand the inevitable onslaught of attacks.

Core tenets of the SSDF

The SSDF can be broken down into some high-level objectives. Let's understand these tenets to solidify our understanding:

- **Proactive security**: The SSDF emphasizes threat modeling, secure design principles, and proactive vulnerability identification to address potential issues early, when they're less costly to fix.

- **Depth of defense**: It reinforces the philosophy of DiD by recommending security controls and safeguards at multiple levels (code, infrastructure, network, monitoring, etc.).

- **Continuous improvement**: The SSDF recognizes that security is a journey, not a destination. Regular testing, vulnerability scanning, patch management, and incident response keep defenses evolving to meet current threats.

- **Adaptability**: While providing structure, the SSDF is designed to adapt to different development methodologies (Agile, Waterfall, etc.) and varying risk profiles of different applications.

Understanding the core tenets is crucial to comprehend where they apply. Now, let's turn our focus to where they can help bolster our security practices.

Benefits of the SSDF

The advantages of aligning your security model with industry standards are numerous. Let's explore some benefits of the SSDF:

- **Reduced risk**: By systemically addressing security throughout the life cycle, the SSDF helps organizations create more inherently resilient software that's harder to compromise.

- **Improved compliance**: The SSDF aligns with numerous security standards, making it easier to demonstrate compliance with regulations and industry mandates.

- **Cost savings**: Addressing vulnerabilities early is far more cost-effective than reacting to breaches and security incidents after the fact.

- **Customer trust**: Demonstrating a commitment to security through frameworks such as the SSDF builds trust with users, leading to a competitive advantage.

The following figure depicts what SSDF covers, and we will go through each phase one by one:

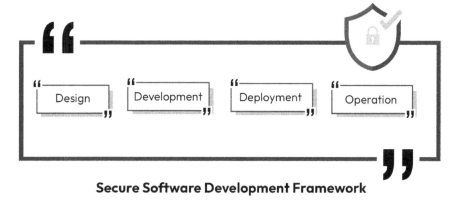

Secure Software Development Framework

Figure 10.1 – The SSDF

We will use this framework as an example to align our security strategy and demonstrate its conversion into a security model. In the rest of this section, we will break it down into different phases and build our skeleton security model from the ground up.

Secure design and requirements

The foundation of any secure system lies in its initial blueprint. The SSDF emphasizes starting with security in mind, not treating it as an afterthought to be patched on later. This philosophy shift is crucial, as early design flaws can become deeply ingrained vulnerabilities that are far more difficult and expensive to address down the line.

Within the *secure design and requirements* phase, the SSDF provides a framework for proactively identifying threats and architecting security into the system's core. This involves processes such as threat modeling, where you methodically step into an attacker's mindset to uncover potential attack vectors. It also encourages defining secure architecture patterns that enforce principles such as least privilege (minimizing access) and compartmentalization (limiting the impact of a breach).

Let's explore some key SSDF recommendations in this area and discuss practical ways to implement them within your organization. We'll cover actionable strategies to translate secure design principles into a tangible and secure development process. By embedding security into the earliest stages, you establish a robust foundation that will pay dividends throughout the system's life cycle.

Establishing secure design principles

The SSDF stresses proactively defining security principles for your organization. These guide architectural choices. Here are some examples:

- **Defense in depth**: This refers to multiple, layered security controls. This is going to be a highlight as we explore this framework.

- **Least privilege**: Components have only the minimum access needed.

- **Fail-safe**: The system defaults to a secure state upon failures.

- **Secure defaults**: Initial configurations are restrictive.

Our goal while defining a security strategy here is to define the core security philosophies that will guide our architectural and development choices. We need to focus on principles that support our organization's specific risk tolerance and security goals.

Performing threat modeling

NIST SP 800-218 emphasizes the methodical identification of threats, mapping them to attack surfaces. This reveals vulnerabilities early. Consider the following steps:

1. **Identify assets**: Identify data, functions, and system components that are valuable.

2. **Decompose the system**: Break it down to understand attack paths.

3. **Identify threats**: Use frameworks such as STRIDE or others introduced in *Chapter 2*.

4. **Rate risks**: Prioritize mitigation based on impact and likelihood.

Basically, we need to proactively think like an attacker, identifying our system's valuable assets, meticulously mapping out how they could be compromised, and focusing on prioritizing the most critical threats for mitigation.

Creating and using secure design patterns

Don't reinvent the wheel! The SSDF suggests building a repository of tested solutions:

- **Authentication and authorization**: Proven ways to manage access securely.

- **Input validation**: Robust patterns to sanitize all user-supplied data.

- **Cryptography**: Standard libraries, avoiding custom encryption schemes.

The idea is to leverage established, proven solutions to common security needs. Focus on patterns for authentication, input validation, and the safe use of cryptography. Sometimes, these will be common libraries developed in-house but shared by all teams.

Defining security requirements

Translate security goals into specific technical requirements alongside functional ones. The following are some examples:

- **Data encryption**: Specify algorithms and key management standards.
- **Password strength**: Mandate complexity and disallow common passwords.
- **Audit logging**: Define which events must be logged for traceability.

We need to translate broad security goals into precise technical specifications. Our focus will stay on requirements that are measurable and testable to ensure they are effectively implemented.

In a nutshell, the SSDF (NIST SP 800-218) recognizes that early focus on security reduces costly rework and inherently builds more resilient systems.

Secure development practices

Even with the best intentions and secure designs, the code itself is where vulnerabilities often creep in. The SSDF's *secure development practices* phase provides guidance on establishing processes, tools, and training that help developers produce inherently safer software. Think of it as translating those secure blueprints into a well-built structure.

The SSDF highlights secure coding techniques to mitigate common vulnerabilities such as injection attacks and **cross-site scripting** (**XSS**). It emphasizes the safe use of libraries and frameworks, encouraging developers to leverage well-vetted components rather than reinventing the wheel. Additionally, the SSDF champions the integration of security tooling directly into the development pipeline, promoting early and frequent vulnerability detection.

We'll explore strategies for empowering developers to write code with security in mind and examine tools such as **static application security testing** (**SAST**) that automate the discovery of potential flaws. The goal is to make secure coding the default, fostering a culture where developers instinctively consider the security implications of each line of code.

Utilizing secure coding practices

The SSDF emphasizes the crucial role of developers in preventing vulnerabilities. Consider focusing on the following:

- **Training on common pitfalls**: Educate developers on things such as the OWASP Top 10, input validation, and injection attacks.

- **Secure coding standards**: Define guidelines for safe language use, error handling, cryptographic libraries, and so on.

- **Code reviews**: Encourage peer reviews or automated linting tools to catch security issues alongside functional bugs.

In a nutshell, we need to invest in security training for our developers, define clear secure coding guidelines, and make code reviews a priority.

Using approved tools and libraries

Leveraging well-tested components reduces the chances of introducing unknown flaws. Here are some points to consider:

- **Vetting external components**: Assess the track record and security practices of third-party libraries before inclusion.

- **Vulnerability management**: Have a process for monitoring and updating dependencies when security issues are discovered.

- **Repository management**: Control where libraries are sourced from to avoid poisoned or tampered-with packages.

In summary, we need to carefully vet third-party code before using it, stay on top of vulnerability reports, and control where our developers get their libraries.

Employing static analysis tools (SAST)

The SSDF promotes integrating security scanning into the development pipeline. Think about the following:

- **Tool selection**: SAST tools vary in language support and the types of vulnerabilities they detect.

- **Integration with CI/CD**: Automate scans with every build or code commit to catch issues early.

- **Managing results**: Prioritize fixing high-severity issues while using the output as a learning opportunity for developers.

To sum up, we need to select the right SAST tools, make them part of our development process, and teach our developers how to use the results to write safer code.

Protecting the code repository

Code is an intellectual property asset and a tempting target. Focus on the following:

- **Access control**: Implement strict permissions on who can modify code and review commit history.

- **Secrets management**: Never store API keys, passwords, and so on directly in code. Use secure storage solutions.

- **Branching and reviews**: Enforce code review processes for changes, especially those to sensitive areas of the code base.

In short, we need to control access to our code base, remove any secrets stored in it, and enforce a strict process for making and reviewing changes.

The SSDF views secure development as a proactive effort requiring training, tools that are integrated into the workflow, and a focus on protecting the source code itself.

Secure deployment and testing

This may start to sound repetitive, but even perfectly written code can become vulnerable if deployed into an insecure environment. The SSDF emphasizes that security doesn't end at development; it extends to the processes and configurations surrounding the running application. This phase focuses on ensuring secure system deployment and implementing rigorous testing to uncover weaknesses that might have slipped through earlier stages.

The SSDF advocates for hardening – locking down systems by removing unnecessary services, applying security patches, and adhering to secure configuration baselines. It also stresses the importance of proactive vulnerability scanning to identify issues in the production environment, even those that may exist in third-party components.

Additionally, the SSDF recommends engaging in penetration testing. This simulates real-world attacks, helping organizations discover gaps in their defenses that other forms of testing might miss.

Let's explore the SSDF best practices for secure deployment and testing. We'll cover strategies for creating hardened environments, automating vulnerability discovery, and proactively seeking out flaws before attackers can exploit them. By addressing security both within the code and in its surrounding ecosystem, you build a system resilient to attacks.

Performing vulnerability scanning

The SSDF stresses continuous vigilance, even in production. Some points to consider are as follows:

- **Regular scans**: Schedule vulnerability scans of deployed systems, covering OS, applications, and network configurations.

- **Tool selection**: Choose scanners that align with your technologies and can detect the latest vulnerabilities.

- **Prioritizing remediation**: Address critical vulnerabilities swiftly and incorporate fixes into future build processes.

Overall, we need to regularly scan our production systems for vulnerabilities, prioritize fixing the worst ones, and learn from them to make our code stronger.

Conducting configuration and change management

Configuration drift is a major security risk. You should focus on the following:

- **Secure baselines**: Develop hardened configurations tailored to your systems and enforce them consistently.

- **Change management**: Document and review all system configuration changes, preventing unauthorized modifications.

- **Configuration as code**: Manage configurations with version control and automation tools for consistency and auditability.

To summarize, we need to treat our system configurations as code, version control them, make changes carefully, and always stick to secure defaults.

Managing the security of external components

Vulnerabilities in third-party code can expose your systems. When protecting a software/system, you should think about the following:

- **Inventory and monitoring**: Maintain an up-to-date list of dependencies and track known vulnerability alerts.

- **Patching processes**: Have a plan for how to analyze, test, and rapidly deploy security updates.

- **Sandboxing as needed**: For high-risk components, consider isolating them to limit potential damage if compromised.

In simple words, we need to know what third-party code we're using, watch it for security problems, and have a fast way to update it when needed.

Conducting penetration testing

The SSDF encourages simulating real attacks to find what other tests might miss. While designing your penetration testing efforts, consider the following aspects:

- **Scoping and planning**: Pentests should have clear goals, be approved in advance, and avoid disruption to production.

- **External expertise**: Often, using third-party pentesters provides specialized skills and a fresh perspective.

- **Remediation focus**: The goal isn't just to find flaws but to learn from them to strengthen defenses overall.

Basically, we need to hire hackers (the good kind) or start to think like them and try to break into our systems so we can find the weak spots before the bad guys do.

The SSDF views secure deployment and testing as ongoing processes, not one-time events. Automation and proactive vulnerability management are crucial to staying ahead of an evolving threat landscape.

Secure operation and maintenance

Once again, it's tempting to think that once a system is deployed, the security work is done. Sadly, this is far from the truth. The SSDF recognizes that the world of cyber threats is dynamic. New vulnerabilities emerge, attackers evolve their tactics, and the systems themselves change over time. This phase focuses on maintaining a security-conscious mindset throughout the operational life cycle.

The SSDF highlights the importance of rigorous patch management. This involves promptly applying updates that address known security flaws. It also stresses the crucial role of robust logging and monitoring. These tools help detect anomalous activity, potentially signaling an attack in progress.

Incident response is another key pillar within the *secure operations and maintenance* phase. The SSDF advocates not just for having a plan in the event of a breach, but also for regularly testing and refining that plan. The goal is to minimize the impact and recover from attacks as swiftly as possible.

Let's explore the SSDF best practices for secure operations. We'll cover strategies for staying ahead of vulnerabilities, effectively responding to incidents, and fostering a culture where security remains a top priority, even in the day-to-day running of systems.

Managing security patches

The SSDF emphasizes the criticality of timely patch deployment to address known vulnerabilities. You should consider the following:

- **Patching sources**: Know where to find security updates for all software you use (OS, frameworks, middleware, etc.).

- **Prioritization**: Assess risk based on the severity of vulnerabilities, your exposure, and the potential for disruption.

- **Testing rigor**: Establish a testing process for patches, especially those impacting critical systems.

In short, we need to know where to get security patches for everything we use, test them quickly, and have a clear plan for rolling them out.

Employing monitoring and logging

The ability to detect attacks early is crucial. Pay attention to the following:

- **What to log**: Define the key events relevant to security (login attempts, configuration changes, etc.).

- **Centralized collection**: Aggregate logs for analysis, preventing attackers from erasing their tracks from individual systems.

- **Alerting and analysis tools**: Use SIEM solutions or automated rules to spot anomalies and generate actionable alerts.

To summarize, we need to know what normal looks like on our systems, watch for anything unusual, and use tools to help us analyze the data.

Establishing incident response capability

It is important to remember that breaches *will* happen. The SSDF stresses being prepared. When designing your security strategy, think about the following:

- **Incident response plan**: Define roles, communication channels, containment procedures, and recovery steps.

- **Practice and review**: Conduct drills to test the plan, identify weaknesses, and update it regularly.

- **Forensics readiness**: Have processes and tools to collect evidence for post-breach investigation and potential legal actions.

In a nutshell, we need to have a detailed plan for what to do when things go wrong, practice it like a fire drill, and be ready to investigate afterward to learn from it.

Conducting security training/awareness

People are your last line of defense, but only if you train your workforce well. Keep security top of mind with the following:

- **User awareness**: Train all personnel to spot phishing and social engineering tactics, and to report suspicious activity.

- **Operational security**: Train those with system access on secure practices, reducing the risk of insider threats.

It's important to remember that security is everyone's job, and to regularly train your people to recognize threats and know how to protect your organization.

The SSDF views security as a dynamic process. Patching, monitoring, incident preparedness, and continuous training are essential to adapt and maintain resilience long after your systems go live. The NIST framework we've discussed has a lot more nuances that we did not get a chance to cover here, so consider referring to the *Further reading* section for a deep dive into the standard.

In this section, we've employed the SSDF/NIST SP 800-218 framework as a guiding structure, outlining comprehensive defense strategies that are suitable for organizational implementation. This approach can be seamlessly extended to other relevant security guidelines and standards within your industry.

A closer examination reveals a systematic approach to constructing a security model that's grounded in industry best practices. Notably, the detailed exploration of individual security controls underscores their significance throughout the book. Our aim has been to equip you with the practical knowledge needed to integrate these insights directly into enhancing your organization's security posture.

Next, we will take a final look at the importance of adaptability in any DiD security framework.

Continuously monitoring and improving security posture

The comforting notion that a system can ever be declared "perfectly safe" is a dangerous illusion; we can only build safer systems. In the realm of cybersecurity, static is synonymous with vulnerability. Adversaries relentlessly hone their tactics, new technologies expose unforeseen cracks in our defenses, and the sheer complexity of modern systems makes it impossible to guarantee perfection. DiD acknowledges this reality, but it doesn't mean we give up.

The key to survival in the digital arms race of a constantly evolving threat landscape lies in continuous monitoring and improvement. Yesterday's cutting-edge security measures become tomorrow's easily bypassed obstacles. To stay ahead of the curve, we must constantly adapt, refine, and innovate our defenses.

This mindset is especially critical given the frightening advancements in attacker capabilities. AI-powered agents/tools automate vulnerability discovery, making exploits faster to create and harder to anticipate. Zero-day attacks strike without warning, highlighting the need for proactive defenses and lightning-fast responses. Moreover, supply chain compromises underscore the risk posed by dependencies that we don't directly control.

This section reiterates the importance of embracing continuous improvement as a core principle of a resilient DiD strategy. We'll explore tools and techniques that enable organizations to stay agile in the face of emerging threats. We'll briefly revisit threat intelligence, behavioral analytics, and deliberate failure injection, all of which aim to identify weaknesses before attackers exploit them.

Ultimately, we'll emphasize building a culture of continuous security. This involves embracing blameless postmortems to learn from failure, fostering proactive vulnerability discovery, and embedding security from the earliest stages of design all the way through to production operations.

Changing the mindset

We began this chapter by acknowledging one of the most intriguing aspects of security: its relativity. Unless we strip away everything of value from our systems, achieving "perfect security" remains an elusive goal – and even then, it renders the software useless. In *Chapter 3*, we discussed how policies and standards can serve as guides for organizations to construct resilient defense strategies, yet these standards are often treated merely as checklists. Earlier, in the section on the SSDF, we demonstrated how a standard can be leveraged to establish a security baseline, emphasizing the crucial aspect of adaptability.

The first step toward continuous improvement involves shattering this illusion of ever achieving a perfect score on a "checklist." DiD itself acknowledges that any single layer might fail, and true resilience comes from continuous improvement.

The goalposts are perpetually shifting. Attackers relentlessly innovate, leveraging the latest tools and AI-driven techniques. Vulnerabilities lurk in the increasing complexity of our software, and even unknown flaws (zero-days) pose immense risk. This isn't a cause for despair, but rather a call for a fundamental shift in our thinking.

Instead of viewing security as a destination, we need to embrace it as an ongoing journey. Defenses that were robust yesterday might be woefully inadequate against the attacks devised tomorrow. We must view our systems with a healthy dose of skepticism, constantly asking, "How might this be broken? What haven't we considered?" This mindset shift is crucial to fostering the proactive improvement that keeps us one step ahead of the adversary.

Building a culture of continuous improvement

The tools and techniques we've discussed in this book – threat intelligence, advanced analytics, and failure injection simulations – are all powerful weapons in your security arsenal. But no technology is a silver bullet. True resilience comes from fostering a culture of continuous improvement within your security team and organization as a whole.

This section delves into practical strategies for nurturing such a culture. Remember, security is everyone's responsibility; it's not just about the security team working tirelessly behind the scenes. Here, we'll explore real-world practices that encourage collaboration, empower individuals to contribute, and foster a shared sense of responsibility for maintaining a robust security posture.

By implementing these practices, we can move beyond a reactive approach to security where we're constantly patching vulnerabilities discovered by attackers. Instead, we can become proactive by constantly learning, refining our defenses, and adapting to the ever-evolving threat landscape.

Now, let's dive into specific techniques that can make continuous improvement a natural part of your security strategy:

- **Embracing failure (the right way)**: Incident response shouldn't be about finding a scapegoat. Instead, implement blameless postmortems. These are focused on systemic analysis – what process fails, gaps in monitoring, or vulnerabilities allowed the incident to escalate? This turns every breach into a valuable, albeit painful, learning opportunity.

- **Gamifying vulnerability discovery**: Make finding security issues fun and rewarding. Internal bug bounty programs, capture-the-flag exercises, and leaderboards for teams with the fastest vulnerability remediation times can incentivize proactive security work. Public recognition of individuals finding significant flaws reinforces positive behavior. Integrating security best practices that positively influence performance reviews at the organization level can be a game changer.

- **Automating early and often**: The earlier a problem is detected, the cheaper and less disruptive the fix. Integrate scanning tools (SAST, DAST, container security, etc.) into the CI/CD pipeline so developers get immediate feedback. Automate routine security tasks such as dependency updates, freeing up your team to focus on higher-level problems and innovation.

- **Chaos engineering as practice**: Don't wait for systems to fail in production to test your resilience. Chaos engineering involves deliberately injecting failures (taking down a service, simulating network issues, etc.) in controlled environments. We discussed this topic in an earlier chapter. It exposes weaknesses, battle-tests incident response processes, and forces teams to design systems that gracefully handle the unexpected.

- **Metrics that motivate**: Shift the focus from "number of vulnerabilities patched" to metrics that reflect real improvement. Track things such as the reduction in time-to-detect and time-to-remediate incidents. Measure the decrease in the severity of vulnerabilities discovered over time (hopefully, by catching them earlier in the SDLC). This demonstrates the tangible impact of security initiatives.

- **Continuous learning of industry best practices**: One core advantage of defenders is the ability to collaborate. While attackers' top priority is to reduce the trust circle, as defenders, you can collaborate across the industry. Investing in learning from the wider security community is advantageous. This is done by reading publications and attending conferences and roundtables.

A continuous improvement culture isn't built on fear, but on empowerment, a shared sense of purpose, and a relentless drive to proactively address risks before they become catastrophes. In cybersecurity, the only true constant is change. Embracing this fluidity is the first step toward building a truly resilient DiD strategy.

Security tomorrow – Sustaining a living DiD

Throughout this book, we've explored the principles and practical implementation of DiD. We've seen how a layered, multi-pronged security strategy offers resilience in a world where attacks are inevitable and absolute security remains an illusion. However, understanding DiD intellectually is only part of the battle. The other part lies in sustaining it as a living, evolving model within your organization.

The reality is that security is often an uphill battle. On one hand, security teams must face friction to get leadership buy-ins inside the organization, and on the other hand, defenders are pitted against an adversary fueled by innovation and financial gain. Attackers only have to find one crack in your armor, while defenders must be constantly vigilant across a sprawling attack surface. A successful breach makes headlines, but the countless thwarted attacks are invisible triumphs.

This asymmetry can be demoralizing. It's easy to fall into a sense of futility or become complacent when attacks seem relentless. Yet, surrender is not an option. The stakes, from financial losses to damaged reputations and even threats to public safety, have never been greater as our reliance on technology grows.

In this closing section, we'll reaffirm why DiD remains a crucial weapon in this fight. We'll discuss the psychological aspects of defense, the need for relentless adaptation, and the role of automation in a future where the sheer volume of threats demands it. We'll leave you with a renewed awareness of the challenges defenders face, and a roadmap for how to not only survive but thrive in this ever-evolving cyber conflict.

The defender's mindset

Security professionals wage a relentless war on an invisible battlefield. Their work often goes unnoticed, except when things go wrong. Understanding the unique mindset required of defenders is crucial to sustaining a resilient DiD approach. Let's go over some common ways of thinking.

Acknowledging the odds

The reality is that cybersecurity is an inherently asymmetrical conflict. Attackers need to find a single weakness, while defenders must strive to protect everything, all the time. A successful attack can undo months, even years, of painstaking security work. It's crucial to accept this imbalance, not as a cause for despair, but as a driving force for vigilance and the need for layered defenses.

This asymmetry is compounded by the sheer dynamism of the threat landscape. Attackers constantly evolve, aided by the rapid advancement of technology. Artificial intelligence and automated tools enable them to probe for weaknesses and launch attacks at an unprecedented scale. The latest zero-day vulnerability might remain undiscovered by the "good folks" for days, weeks, or even longer.

Understanding these odds underscores why DiD is non-negotiable. It acknowledges the impossibility of guaranteeing that any one safeguard will hold forever. By building multiple layers of protection, we increase the cost, time, and resources required for an attacker to succeed. The goal is to make a breach so difficult and frustrating that they'll seek an easier target.

Embracing imperfection

Perfection is the enemy of good in the world of security. Despite the best efforts, breaches and vulnerabilities are inevitable. The key lies in building systems designed for resilience, assuming something will fail. This involves swift detection, containment to minimize damage, and, crucially, a relentless drive to learn and improve defenses from each incident.

A fixation on perfection can be paralyzing within cybersecurity. The fear of failure can lead to excessive delays as teams strive for the unattainable ideal of an "unbreakable" system. Ironically, this often backfires, as security enhancements get stuck in endless analysis while vulnerabilities remain unpatched. This is where embracing imperfection comes in.

Fighting complacency

The endless news cycle of breaches can breed cynicism or a sense of futility. However, every averted attack and every vulnerability patched proactively is a victory. A strong defender's mindset requires celebrating these small wins and fostering a team culture focused on continuous improvement. This means taking pride in the relentless, often unseen, efforts that keep systems safe day after day. We don't hesitate to pour endless resources into improving our software; it's time to stop treating security as a deadline.

The automation imperative

The relentless march of technology generates overwhelming volumes of data; logs, alerts, network traffic patterns, indicators of compromise, and more. Relying solely on human analysts to sift through this avalanche of information is a recipe for burnout and missed critical threats. This is where automation becomes indispensable.

The challenge of scale

Even a modestly sized organization can generate millions of security events per day. Expecting humans to manually identify the truly dangerous anomalies within this haystack is neither realistic nor scalable. Automated tools can process vast amounts of data in real time, tirelessly looking for patterns that might signal an attack.

In reality, this volume problem is magnified by the distributed nature of modern systems. Applications span on-premises servers, cloud infrastructure, and countless user devices. Each of these components generates data relevant to security. Attempting to make sense of it all manually creates bottlenecks and blind spots where attackers can thrive.

Moreover, the speed and sophistication of attacks intensify the issue. Automated malware can spread at alarming rates, and attackers leverage AI to adapt their techniques. Human response times simply can't compete. Automation offers a chance to fight back at a pace that aligns with the modern threat landscape.

The rise of AI-powered security

AI/ML techniques are revolutionizing cybersecurity on both sides. Attackers have already started using modern techniques to launch large-scale exploits. Defenders can't stay far behind. AI-powered tools can learn to distinguish normal system behavior from subtle deviations that indicate an intrusion attempt. They can automatically correlate events across different systems, potentially revealing a sophisticated attack's early stages. AI can even be used to prioritize threats and suggest remediation actions, aiding security teams.

The human-machine partnership

It's crucial to stress that automation doesn't replace human expertise. Instead, it aids security analysts by filtering out the noise and elevating critical threats for investigation. This allows skilled professionals to focus on high-level analysis, proactive threat hunting, and refining the automated defenses themselves. Think of it as equipping defenders with ever-sharper tools, not seeking to eliminate the need for their unique skills and judgment.

More importantly, the relationship between human experts and automation must be collaborative. Security professionals train and refine the AI models, ensuring they remain relevant to the specific threats faced by their organization. They're the ones with the intuition and deep contextual knowledge to interpret ambiguous alerts and guide further investigation.

Automation, in turn, frees up those same experts from mundane and repetitive tasks. It allows them to focus on higher-order challenges such as strategizing how to outsmart attackers, proactively identifying new vulnerabilities within the organization's systems, and continuously evolving the DiD model to stay ahead of emerging threats.

DiD as an organizational value

DiD cannot remain the sole domain of security teams. While specialized expertise is essential, true resilience is only achievable when security is woven into the fabric of an organization and embraced as a shared value at every level. This transformation requires a paradigm shift in how we approach technology development.

Developers must view security as an integral feature of their code, not an afterthought tacked on before release. This means proactive use of secure coding practices, participation in threat modeling, and seamlessly integrating security testing into their workflows.

Leadership must champion security initiatives, fostering a culture where addressing vulnerabilities is prioritized alongside new functionality. This involves allocating budget, time, and resources to proactive security measures, rather than viewing them as costs to be minimized.

Every employee, from executives to end users, has a role to play. Basic security hygiene such as strong passwords, skepticism toward phishing attempts, and the responsible use of sensitive data is the collective first line of defense. Organizations must invest in continuous security awareness training, tailored to different roles, to make this second nature.

The stakes are too high for us to approach cybersecurity casually. By embracing DiD as an enduring organizational value, we can build a future where technology empowers and enriches our lives, not one where we perpetually live in fear of the next devastating breach. Let's commit to the continuous improvement, collaboration, and vigilance needed to make this shared vision of resilient systems a reality.

Summary

We initiated this chapter by revisiting a concept cultivated throughout the book: the absence of a "perfectly safe" system, but rather the pursuit of safer systems. Fully embracing this reality is paramount when crafting any defense strategy, fostering a mindset of vigilance within security teams. Next, we delved into a recent NIST specification, leveraging it to incrementally construct a DiD strategy. The fundamental objective was to provide you with hands-on experience in implementing a framework to instigate organizational changes.

We divided the SSDF into four phases, offering a high-level examination of each. It's essential to recognize that this breakdown served as an illustrative example and doesn't encompass the entirety of the framework. Among the framework's topics, one significant aspect discussed was the fluidity of a security program. Nearly every chapter underscored the importance of continuous monitoring and enhancement, treating defense as an ongoing process characterized by rapid updates.

The world of cybersecurity is in constant flux. Adversaries harness relentless innovation, seeking to exploit new technologies and unforeseen vulnerabilities faster than ever before. In this dynamic landscape, DiD remains a vital beacon. It guides us to build multi-layered defenses that acknowledge the inevitability of individual failures, focus on resilience, and force attackers to overcome multiple hurdles.

Fortifying our systems against tomorrow's attacks requires a fundamental shift. Security can no longer reside solely within specialized teams. Instead, it must become an organizational mindset embraced by developers, leaders, and everyday users. This involves vigilance, continuous improvement, intelligent automation, and a relentless drive to adapt and stay ahead of those who seek to undermine the technology we increasingly depend upon.

Throughout this exploration of DiD, we've journeyed from fundamental security concepts to real-world strategies for building resilient systems. We've embraced the reality that absolute security is a myth, instead focusing on layers of protection, continuous monitoring, and swift response to minimize the impact of inevitable breaches. This book serves as a roadmap, equipping you with the knowledge and principles to make DiD a cornerstone of your security strategy, and empowering you to face the challenges of the modern cyber threat landscape with confidence and proactive determination.

Key takeaways

- DiD needs to be part of any modern security model.
- DiD is a mindset of resilience that acknowledges the inevitability of security failures.
- Defenders face an uphill battle against well-funded and innovative adversaries.
- Security requires continuous adaptation to counter the ever-evolving techniques of attackers.
- DiD is an organizational responsibility, requiring vigilance from developers to executives.
- A proactive, collaborative, and relentless focus is needed to build resilient systems for the future.

Congratulations on completing this book. DiD is a crucial way of thinking about modern security. In this book, we covered a wide range of topics and discussed how DiD relies on security-first principles. Next time you're thinking of designing a secure system, ask yourself how you can leverage a layered approach to keep your customer information safe.

Further reading

To learn more about the topics that were covered in this chapter, take a look at the following resources:

- [1] Automated AI agents: `https://arxiv.org/html/2402.06664v1`

- [2] State of Nature Frankenstein: `https://www.bartleby.com/essay/State-Of-Nature-Frankenstein-5C9EE2BD4005A8FB`

- [3] Blind SQLi: `https://secgroup.dais.unive.it/wp-content/uploads/2018/12/Side-channel-attacks.pdf`

- [4] Spectre and Meltdown: `https://meltdownattack.com/`

- [5] Speculative execution: `https://en.wikipedia.org/wiki/Speculative_execution`

- [6] NIST SSDF: `https://nvlpubs.nist.gov/nistpubs/SpecialPublications/NIST.SP.800-218.pdf`

- [7] Executive Order 14028: `https://www.whitehouse.gov/briefing-room/presidential-actions/2021/05/12/executive-order-on-improving-the-nations-cybersecurity/`

Index

Symbols

Y

Other Books You May Enjoy

If you enjoyed this book, you may be interested in these other books by Packt:

Mastering Cloud Security Posture Management (CSPM)

Qamar Nomani

ISBN: 978-1-83763-840-6

- Find out how to deploy and onboard cloud accounts using CSPM tools
- Understand security posture aspects such as the dashboard, asset inventory, and risks
- Explore the Kusto Query Language (KQL) and write threat hunting queries
- Explore security recommendations and operational best practices
- Get to grips with vulnerability, patch, and compliance management, and governance
- Familiarize yourself with security alerts, monitoring, and workload protection best practices
- Manage IaC scan policies and learn how to handle exceptions

Endpoint Detection and Response Essentials

Guven Boyraz

ISBN: 978-1-83546-326-0

- Gain insight into current cybersecurity threats targeting endpoints
- Understand why antivirus solutions are no longer sufficient for robust security
- Explore popular EDR/XDR tools and their implementation
- Master the integration of EDR tools into your security operations
- Uncover evasion techniques employed by hackers in the EDR/XDR context
- Get hands-on experience utilizing DNS logs for endpoint defense
- Apply effective endpoint hardening techniques within your organization

Packt is searching for authors like you

If you're interested in becoming an author for Packt, please visit authors.packtpub.com and apply today. We have worked with thousands of developers and tech professionals, just like you, to help them share their insight with the global tech community. You can make a general application, apply for a specific hot topic that we are recruiting an author for, or submit your own idea.

Share Your Thoughts

Now you've finished *The Complete Guide to Defense in Depth*, we'd love to hear your thoughts! Scan the QR code below to go straight to the Amazon review page for this book and share your feedback or leave a review on the site that you purchased it from.

https://packt.link/r/1835468268

Your review is important to us and the tech community and will help us make sure we're delivering excellent quality content.

Download a free PDF copy of this book

Thanks for purchasing this book!

Do you like to read on the go but are unable to carry your print books everywhere?

Is your eBook purchase not compatible with the device of your choice?

Don't worry, now with every Packt book you get a DRM-free PDF version of that book at no cost.

Read anywhere, any place, on any device. Search, copy, and paste code from your favorite technical books directly into your application.

The perks don't stop there, you can get exclusive access to discounts, newsletters, and great free content in your inbox daily

Follow these simple steps to get the benefits:

1. Scan the QR code or visit the link below

https://packt.link/free-ebook/9781835468265

2. Submit your proof of purchase
3. That's it! We'll send your free PDF and other benefits to your email directly